Reliability of Geotechnical Structures

Reliability of Geotechnical Structures

Editor

Davin Tuft

scitus
academics

Reliability of Geotechnical Structures
Edited by **Davin Tuft**

ISBN: 978-1-68117-684-0
Library of Congress Control Number: 2016958812

www.scitusacademics.com

Printed in United States of America on Acid Free Paper

Preface

The structural design based on reliability is much more developed. Development and evaluation strategies based on reliability are used to make design and operational decisions that achieve specific target reliability levels. Since modern structures require much more critical and complex designs, the need for precise approaches to assess uncertainties in loads, material properties, geometry, manufacturing processes and operating environments has significantly increased. Reliability assessment techniques help to create safe designs and precisely identify where important contributing factors of uncertainty occur in structural systems or where further research, testing and quality control could improve the structure's safety and efficiency. The book provides an overview of how reliability can play a complementary role within prevailing geotechnical standards in situations where there are measured past experience and data for limited site-specific data to be supplemented by objective regional data and subjective judgment from comparable sites elsewhere.

This book presents all the essential information for a non-specialist to realistically and robustly compute and understand the reliability index and threat of geotechnical constructions. It will be a valuable reference for graduate and postgraduate students studying structural reliability, probabilistic analysis and uncertainty optimization, as well as engineers, researchers and technical managers dealing with theoretical fundamentals, computer implementations and probabilistic analysis applications.

Table of Contents

1

CHAPTER

A Cognitive Look at Geotechnical Earthquake Engineering: Understanding the Multidimensionality of the Phenomena

Silvia Garcia

[1] Geotechnical Department, Institute of Engineering, National University of Mexico, Mexico

1. INTRODUCTION

Essentially, disasters are human-made. For a catastrophic event, whether precipitated by natural phenomena or human activities, assumes the state of a disaster when the community or society affected fails to cope. Earthquake hazards themselves do not necessarily lead to disasters, however intense, inevitable or unpredictable, translate to disasters only to the extent that the population is unprepared to respond, unable to deal with, and, consequently, severely affected. Seismic disasters could, in fact, be reduced if not prevented. With today's advancements in science and technology, including early warning and forecasting of the natural phenomena, together with innovative approaches and strategies for enhancing local capacities, the impact of earthquake hazards somehow could be predicted and mitigated, its detrimental effects on populations reduced, and the communities adequately protected.

After each major earthquake, it has been concluded that the experienced ground motions were not expected and soil behavior and soil-structure interaction were not properly predicted. Failures, associated to inadequate design/construction and to lack of phenomena comprehension, obligate further code reinforcement and research. This scenario will be repeated after each earthquake. To overcome this issue, *Earthquake Engineering* should change its views on the present methodologies and techniques toward more scientific, doable, affordable, robust and adaptable solutions.

A competent modeling of engineering systems, when they are affected by seismic activity, poses many difficult challenges. Any representation designed for reasoning about models of such systems has to be flexible enough to handle various degrees of complexity and uncertainty, and at the same time be sufficiently powerful to deal with situations in which the input signal may or may not be controllable.Mathematically-based models are developed using scientific theories and concepts that just apply to particular conditions. Thus, the core of the model comes from assumptions that for complex systems usually lead to simplifications (perhaps oversimplifications) of the problem phenomena. It is fair to argue that the representativeness of a particular theoretical model largely depends on the degree of comprehension the developer has on the behavior of the actual engineering problem. Predicting natural-phenomena characteristics like those of earthquakes, and thereupon their potential effects at particular sites, certainly belong to a class of problems we do not fully understand. Accordingly, analytical modeling often becomes the bottleneck in the development of more accurate procedures. As a consequence, a strong demand for advanced modeling an identification schemes arises.

Cognitive Computing CC technologies have provided us with a unique opportunity to establish coherent seismic analysis environments in which uncertainty and partial data-knowledge are systematically handled. By seamlessly combining learning, adaptation, evolution, and fuzziness, CC complements current engineering approaches allowing us develop a more comprehensive and unified framework to the effective management of earthquake phenomena. Each CC algorithm has well-defined labels and could usually be identified with specific scientific communities. Lately, as we improved our understanding of these algorithms' strengths and weaknesses, we began to leverage their best features and developed hybrid algorithms that indicate a new trend of co-existence and integration between many scientific communities to solve a specific task.

In this chapter geotechnical aspects of earthquake engineering under a cognitive examination are covered. Geotechnical earthquake engineering, an area that deals with the design and construction of projects in order to resist the effect of earthquakes, requires an understanding of geology, seismology and earthquake engineering. Furthermore, practice of geotechnical earthquake engineering also requires consideration of social, economic and political factors. Via the development of cognitive interpretations of selected topics: i) spatial variation of soil dynamic properties, ii) attenuation laws for rock sites (seismic input), iii) generation of artificial-motion time histories, iv) effects of local site conditions (site effects), and iv) evaluation of liquefaction susceptibility, CC techniques (Neural Networks NNs, Fuzzy Logic FL and Genetic Algorithms GAs) are presented as appealing alternatives for integrated data-driven and theoretical procedures to generate reliable seismic models.

2. GEOTECHNICAL EARTHQUAKE HAZARDS

The author is well aware that standards for geotechnical seismic design are under development worldwide. While there is no need to "reinvent the wheel" there is a requirement to adapt such initiatives to fit the emerging safety philosophy and demands. This investigation also strongly endorses the view that "guidelines" are far more desirable than "codes" or "standards" disseminated all over seismic regions. Flexibility in approach is a key ingredient of geotechnical engineering and the cognitive technology in this area is rapidly advancing.The science and practice of geotechnical earthquake engineering is far from mature and need to be expanded and revised periodically in coming years. It is important that readers and users of the computational models presented here familiarize themselves with the latest advances and amend the recommendations herein appropriately.

This document is not intended to be a detailed treatise of latest research in geotechnical earthquake engineering, but to provide sound guidelines to support rational cognitive approaches. While every effort has been made to make the material useful in a wider range of applications, applicability of the material is a matter for the user to judge. The main aim of this guidance document is to promote consistency of cognitive approach to everyday situations and, thus, improve geotechnical-earthquake aspects of the performance of the built safe-environment.

2.1. A "SOFT" INTERPRETATION OF GROUND MOTIONS

After a sudden rupture of the earth's crust (caused by accumulating stresses, elastic strain-energy) a certain amount of energy radiates from the rupture as seismic waves. These waves are attenuated, refracted, and reflected as they travel through the earth, eventually reaching the surface where they cause ground shaking. The principal geotechnical hazards associated with this event are fault rupture, ground shaking, liquefaction and lateral spreading, and landsliding. Ground shaking is one of the principal seismic hazards that causes extensive damage to the built environment and failure of engineering systems over large areas. Earthquake loads and their effects on structures are directly related to the intensity and duration of ground shaking. Similarly, the level of ground deformation, damage to earth structures and ground failures are closely related to the severity of ground shaking.

In engineering evaluations, three characteristics of ground shaking are typically considered: i) the amplitude, ii) frequency content and iii) significant duration of shaking (time over which the ground motion has relatively significant amplitudes).These characteristics of the ground motion at a given site are affected by numerous complex factors such as the source mechanism, earthquake magnitude, rupture directivity, propagation path of seismic waves, source distance and effects of local soil conditions. There are many unknowns and uncertainties associated with these issues which in turn result in significant uncertainties regarding the characteristics of the ground motion and earthquake loads.

If the random nature of response to earthquakes (aleatory uncertainty) cannot be avoided [1,2], it is our limited knowledge about the patterns between seismic events and their manifestations - ground motions- at a site (epistemic uncertainty) that must be improved thorough more scientific seismic analyses. A strategic factor in seismic hazard analysis is the ground motion model or attenuation relation. These attenuation relationships has been developed based on magnitude, distance and site category, however, there is a tendency to incorporate other parameters, which are now known to be significant, as the tectonic environment, style of faulting and the effects of topography, deep basin edges and rupture directivity. These distinctions are recognized in North America, Japan and New Zealand [3-6], but ignored in most other regions of the world [7]. Despite recorded data suggest that ground motions depend, in a significant way, on these aspects, these inclusions did not have had a remarkable effect on the predictions confidence and the geotechnical earthquake engineer prefers the basic and clear-cut approximations on those that demand a *blind* use of coefficients or an intricate determination of soil/fault conditions.

A key practice in current aseismic design is to develop design spectrum compatible time histories. This development entails the modification of a time history so that its response spectrum matches within a prescribed tolerance level, the target design spectrum. In such matching it is important to retain the phase characteristics of the selected ground motion time history. Many of the techniques used to develop compatible motions do not retain the phase [8]. The response spectrum alone does not adequately characterize specific-fault ground motion. Near-fault ground motions must be characterized by a long period pulse of strong motion of a fairly brief duration rather than the stochastic process of long duration that characterizes more distant ground motions. Spectrum compatible with these specific motions will not have these characteristics unless the basic motion being modified to ensure compatibility has these effects included. Spectral compatible motions could match the entire spectrum but the problem arises on finding a "real" earthquake time series that match the specific nature of ground motion. For nonlinear analysis of structures, spectrum compatible motions should also correspond to the particular energy input [9], for this reason, designers should be cautious about using spectrum compatible motions when estimating the displacements of embankment dams and earth structures under strong shaking, if the acceptable performance of these structures is specified by criteria based on tolerable displacements.

Another important seismic phenomenon is the liquefaction. Liquefaction is associated with significant loss of stiffness and strength in the shaken soil and consequent large ground deformation. Particularly damaging for engineering structures are cyclic ground movements during the period of shaking and excessive residual deformations such as settlements of the ground and lateral spreads. Ground surface disruption including surface cracking, dislocation, ground distortion, slumping and permanent deformations, large settlements and lateral spreads are commonly observed at liquefied sites. In sloping ground and backfills behind retaining structures in waterfront areas, liquefaction often results in large permanent ground displacements in the down-slope direction or towards waterways (lateral spreads). Dams, embankments and sloping ground near riverbanks where certain shear strength is required for stability under gravity loads are particularly prone to such failures. Clay soils may also suffer some loss of strength during shaking but are not subject to boils and other "classic" liquefaction phenomena. For intermediate soils, the transition from "sand like" to "clay-like" behavior depends primarily on whether the soil is a matrix of coarse grains with fines contained within the pores or a matrix of plastic fines with coarse grained "filler". Recent papers by Boulanger and Idriss [10, 11] are helpful in clarifying issues surrounding the liquefaction and strain softening of different soil types during strong ground shaking. Engineering judgment based on good quality investigations and data interpretation should be used for classifying such soils as liquefiable or non-liquefiable.

Procedures for evaluating liquefaction, potential and induced lateral spread, have been studied by many engineering committees around the world. The objective has been to review research and field experience on liquefaction and recommended standards for practice. Youd and Idriss [12] findings and the liquefaction-resistance chart proposed by Seed et al. [13] in 1985, stay as standards for practice. They have been slightly modified to adjust new registered input-output conditions and there is a strong tendency to recommend i) the adoption of the cone penetration test CPT, standard penetration test SPT or the shear wave velocities for describing the *in situ* soil conditions [14] and ii) the modification of magnitude factors used to convert the critical stress ratios from the liquefaction assessment charts (usually developed for M7:5) to those appropriate for earthquakes of diverse magnitudes [12, 15].

3. COGNITIVE COMPUTING

Cognitive Computing CC as a discipline in a narrow sense, is an application of computers to solve a given computational problem by imperative instructions; while in a broad sense, it is a process to implement the instructive intelligence by a system that transfers a set of given information or instructions into expected behaviors. According to theories of cognitive informatics [16-18], computing technologies and systems may be classified into the categories of imperative, autonomic, and cognitive from the bottom up. Imperative computing is a traditional and passive technology based on stored-program controlled behaviors for data processing [19-24]. An autonomic computing is goal-driven and self-decision-driven technologies that do not rely on instructive and procedural information [25-28]. Cognitive computing is more intelligent technologies beyond imperative and autonomic computing, which embodies major natural intelligence behaviors of the brain such as thinking, inference, learning, and perceptions.

Cognitive computing is an emerging paradigm of intelligent computing methodologies and systems, which implements computational intelligence by autonomous inferences and perceptions mimicking the mechanisms of the brain. This section presents a brief description on the theoretical framework and architectural techniques of cognitive computing beyond conventional imperative and autonomic computing technologies. Cognitive models are explored on the basis of the latest advances in applying computational intelligence. These applications of cognitive computing are described from the aspects of cognitive search engines, which demonstrate how machine and computational intelligence technologies can drive us toward autonomous knowledge processing.

3.1. COMPUTATIONAL INTELLIGENCE: SOFT COMPUTING TECHNOLOGIES

The *computational intelligence* is a synergistic integration of essentially three computing paradigms, viz. neural networks, fuzzy logic and evolutionary computation entailing probabilistic reasoning (belief networks, genetic algorithms and chaotic systems) [29]. This synergism provides a framework for flexible information processing applications designed to operate in the real world and is commonly called *Soft Computing SC* [30]. Soft computing technologies are robust by design, and operate by trading off precision for tractability. Since they can handle uncertainty with ease, they conform better to real world situations and provide lower cost solutions.

The three components of soft computing differ from one another in more than one way. Neural networks operate in a numeric framework, and are well known for their learning and generalization capabilities. Fuzzy systems [31] operate in a linguistic framework, and their strength lies in their capability to handle linguistic information and perform approximate reasoning. The evolutionary computation techniques provide powerful search and optimization methodologies. All the three facets of soft computing differ from one another in their time scales of operation and in the extent to which they embed *a priori* knowledge.

Figure 1 shows a general structure of Soft Computing technology. The following main components of SC are known by now: fuzzy logic FL, neural networks NN, probabilistic reasoning PR, genetic algorithms GA, and chaos theory ChT (Figure 1). In SC FL is mainly concerned with imprecision and approximate reasoning, NN with learning, PR with uncertainty and propagation of belief, GA with global optimization and search and ChT with nonlinear dynamics. Each of these computational paradigms (emerging reasoning technologies) provides us with complementary reasoning and searching methods to solve complex, real-world problems. In large scope, FL, NN, PR, and GA are complementary rather that competitive [32-34]. The interrelations between the components of SC, shown in Figure 1, make the theoretical foundation of Hybrid Intelligent Systems. As noted by L. Zadeh: "... the term hybrid intelligent systems is gaining currency as a descriptor of systems in which FL, NC, and PR are used in combination. In my view, hybrid intelligent systems are the wave of the future" [35]. The use of Hybrid Intelligent Systems are leading to the development of numerous manufacturing system, multimedia system, intelligent robots, trading systems, which exhibits a high level of MIQ (machine intelligence quotient).

3.1.1. COMPARATIVE CHARACTERISTICS OF SC TOOLS

The constituents of SC can be used independently (fuzzy computing, neural computing, evolutionary computing etc.), and more often in combination [36, 37, 38- 40, 41]. Based on independent use of the constituents of Soft Computing, fuzzy technology, neural technology, chaos technology and others have been recently applied as emerging technologies to both industrial and non-industrial areas.

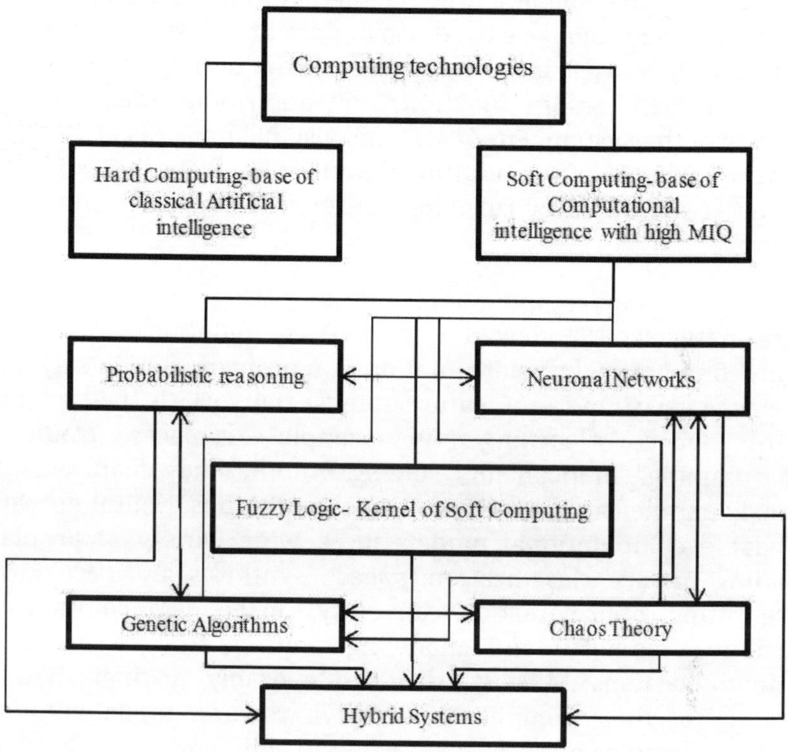

FIGURE 1. Soft Computing Components

Fuzzy logic is the leading constituent of Soft Computing. In Soft Computing, fuzzy logic plays a unique role. FL serves to provide a methodology for computing [36]. It has been successfully applied to many industrial spheres, robotics, complex decision making and diagnosis, data compression, and many other areas. To design a system processor for handling knowledge represented in a linguistic or uncertain numerical form we need a fuzzy model of the system. Fuzzy sets can be used as a universal approximator, which is very important for modeling unknown objects. If an operator cannot tell linguistically what kind of action he or she takes in a specific situation, then it is quite useful to model his/her control actions using numerical data. However, fuzzy logic in its so called *pure form* is not always useful for easily constructing intelligent systems. For example, when a designer does not have sufficient prior information (knowledge) about the system, development of acceptable fuzzy rule base becomes impossible. As the complexity of the system increases, it becomes difficult to specify a correct set of rules and membership functions for describing adequately the behavior of the system. Fuzzy systems also have the disadvantage of not being able to extract additional knowledge from the experience and correcting the fuzzy rules for improving the performance of the system.

Another important component of Soft Computing is neural networks. Neural networks NN viewed as parallel computational models, are parallel fine-grained implementation of non-linear static or dynamic systems. A very important feature of these networks is their adaptive nature, where "learning by example" replaces traditional "programming" in problems solving. Another key feature is the intrinsic parallelism that allows fast computations. Neural networks are viable computational models for a wide variety of problems including pattern classification, speech synthesis and recognition, curve fitting, approximation capability, image data compression, associative memory, and modeling and control of non-linear unknown systems [42, 43]. NN are favorably distinguished for efficiency of their computations and hardware implementations. Another advantage of NN is generalization ability, which is the ability to classify correctly new patterns. A significant disadvantage of NN is their poor interpretability. One of the main criticisms addressed to neural networks concerns their black box nature [35].

Evolutionary Computing EC is a revolutionary approach to optimization. One part of EC—genetic algorithms—are algorithms for global optimization. Genetic algorithms GAs are based on the mechanisms of natural selection and genetics [44]. One advantage of genetic algorithms is that they effectively implement parallel multi-criteria search. The mechanism of genetic algorithms is simple. Simplicity of operations and powerful computational effect are the two main advantages of genetic algorithms. The disadvantages are the problem of convergence and the absence of strong theoretical foundation. The requirement of coding the domain of the real variables' into bit strings also seems to be a drawback of genetic algorithms. It should be also noted that the computational speed of genetic algorithms is low.

Because in this investigation PR and ChT are not exploited, they are not going to be explained. For the interested reader [41] is recommended. Table 1 presents the comparative characteristics of the components of Soft Computing. For each component of Soft Computing there is a specific class of problems, where the use of other components is inadequate.

3.1.2. INTELLIGENT COMBINATIONS OF SC

As it was shown above, the components of SC complement each other, rather than compete. It becomes clear that FL, NC and GA are more effective when used in combinations. Lack of interpretability of neural networks and poor learning capability of fuzzy systems are similar problems that limit the application of these tools. Neurofuzzy systems are hybrid systems which try to solve this problem by combining the learning capability of connectionist models with the interpretability property of fuzzy systems. As it was noted above, in case of dynamic work environment, the automatic knowledge base correction in fuzzy systems becomes necessary. On the other hand, artificial neural networks are successfully used in problems connected to knowledge acquisition using learning by examples with the required degree of precision.

Incorporating neural networks in fuzzy systems for fuzzification, construction of fuzzy rules, optimization and adaptation of fuzzy knowledge base and implementation of fuzzy reasoning is the essence of the Neurofuzzy approach.

TABLE 1. Central characteristics of Soft Computing technologies

	Fuzzy Sets	Artificial Neural Networks	Evolutionary Computing, GA	Probabilistic Reasoning	Chaotic computing
Weaknesses	•Knowledge acquisition •Learning	•Black Box interpretability	•Coding •Computational speed	•Limitation of the axioms of Probability Theory •Lack of complete knowledge •Copmputational complexity	•Computational complexity •Chaos identification complexity
Strengths	•Interpretability •Transparency •Plausibility •Graduality •Modeling •Reasoning •Tolerance to imprecision	•Learning •Adaptation •Fault tolerance •Curve fiting •Generalization ability •Approximation ability	Computational efficiency •Global optimization	•Rigorous framework •Well understanding	•Nonlinear dynamics simulation •Discovering chaos in observed data (with noise) •Determinig the predictability •Prediction startegies formulation

The combination of genetic algorithms with neural networks yields promising results as well. It is known that one of main problems in development of artificial neural systems is selection of a suitable learning method for tuning the parameters of a neural network (weights, thresholds, and structure). The most known algorithm is the "error back propagation" algorithm. Unfortunately, there are some difficulties with "back propagation". First, the effectiveness of the learning considerably depends on initial set of weights, which are generated randomly. Second, the "back propagation", like any other gradient-based method, does not avoid local minima. Third, if the learning rate is too slow, it requires too much time to find the solution. If, on the other hand, the learning rate is too high it can generate oscillations around the desired point in the weight space. Fourth, "back propagation" requires the activation functions to be differentiable. This condition does not hold for many types of neural networks. Genetic algorithms used for solving many optimization problems when the "strong" methods fail to find appropriate solution, can be successfully applied for learning neural networks, because they are free of the above drawbacks.

The models of artificial neurons, which use linear, threshold, sigmoidal and other transfer functions, are effective for neural computing. However, it should be noted that such models are very simplified. For example, reaction of a biological axon is chaotic even if the input is periodical. In this aspect the more adequate model of neurons seems to be chaotic. Model of a chaotic neuron can be used as an element of chaotic neural networks. The more adequate results can be obtained if using fuzzy chaotic neural networks, which are closer to biological computation. Fuzzy systems with If-Then rules can model non-linear dynamic systems and capture chaotic attractors easily and accurately. Combination of Fuzzy Logic and Chaos Theory gives us useful tool for building system's chaotic behavior into rule structure. Identification of chaos allows us to determine predicting strategies. If we use a Neural Network Predictor for predicting the system's behavior, the parameters of the strange attractor (in particular fractal dimension) tell us how much data are necessary to train the neural network. The combination of Neurocomputing and Chaotic computing technologies can be very helpful for prediction and control.

The cooperation between these formalisms gives a useful tool for modeling and reasoning under uncertainty in complicated real-world problems. Such cooperation is of particular importance for constructing perception-based intelligent information systems. We hope that the mentioned intelligent combinations will develop further, and the new ones will be proposed. These SC paradigms will form the basis for creation and development of Computational Intelligence.

4. COGNITIVE MODELS OF GROUND MOTIONS

The existence of numerous databases in the field of civil engineering, and in particular in the field of geotechnical earthquake, has opened new research lines through the introduction of analysis based on soft computing. Three methods are mainly applied in this emerging field: the ones based on the Neural Networks NN, the ones created using Fuzzy Sets FS theory and the ones developed from the Evolutionary Computation [45].

The SC hybrids used in this investigation are directed to tasks of prediction (classification and/or regression). The central objective is obtaining numerical and/or categorical values that mimic input-output conditions from experimentation and in situ measurements and then, through the recorded data and accumulated experience, predict future behaviors. The examples presented herein have been developed by an engineering committee that works for generating useful guidance to geotechnical practitioners with geotechnical seismic design. This effort could help to minimize the perceived significant and undesirable variability within geotechnical earthquake practice. Some urgency in producing the alternative guidelines was seen, after the most recent earthquakes disasters, as being necessary with a desire to avoid a long and protracted process. To this end, a two stage approach was suggested with the first stage being a cognitive interpretation of well-known procedures with appropriate factors for geotechnical design, and a posterior step identifying the relevant philosophy for a new geotechnical seismic design.

4.1. SPATIAL VARIATION OF SOIL DYNAMIC PROPERTIES

The spatial variability of subsoil properties constitutes a major challenge in both the design and construction phases of most geo-engineering projects. Subsoil investigation is an imperative step in any civil engineering project. The purpose of an exploratory investigation is to infer accurate information about actual soil and rock conditions at the site. Soil exploration, testing, evaluation, and field observation are well-established and routine procedures that, if carried out conscientiously, will invariably lead to good engineering design and construction. It is impossible to determine the optimum spacing of borings before an investigation begins because the spacing depends not only on type of structure but also on uniformity or regularity of encountered soil deposits. Even the most detail soil maps are not efficient enough for predicting a specific soil property because it changes from place to place, even for the same soil type. Consequently interpolation techniques have been extensively exploited. The most commonly used methods are kriging and co-kriging but for better estimations they require a great number of measurements available for each soil type, what is generally impossible.

Based on the high cost of collecting soil attribute data at many locations across landscape, new interpolation methods must be tested in order to improve the estimation of soil properties. The integration of GIS and Soft Computing SC offers a potential mechanism to lower the cost of analysis of geotechnical information by reducing the amount of time spent understanding data. Applying GIS to large sites, where historical data can be organized to develop multiple databases for analytical and stratigraphic interpretation, originates the establishment of spatial/chronological efficient methodologies for interpreting properties (soil exploration) and behaviors (in situ measured). GIS-SC modeling/simulation of natural systems represents a new methodology for building predictive models, in this investigation NN and GAs, nonparametric cognitive methods, are used to analyze physical, mechanical and geometrical parameters in a geographical context. This kind of spatial analysis can handle uncertain, vague and incomplete/redundant data when modeling intricate relationships between multiple variables. This means that a NN has not constraints about the spacing (minimum distance) between the drill holes used for building (training) the SC model. The NNs-GAs acts as computerized architectures that can approximate nonlinear functions of several variables, this scheme represent the relations between the spatial patterns of the stratigraphy without restrictive assumptions or excessive geometrical and physical simplifications.

The geotechnical data requirements (geo-referenced properties) for an easy integration of the SC technologies are explained through an application example: a geo-referenced three-dimensional model of the soils underlying Mexico City. The classification/prediction criterion for this very complex urban area is established according to two variables: the cone penetration resistance qcqc (mechanical property) and the shear wave velocity VsVs (dynamic property). The expected result is a 3D-model of the soils underlying the city area that would eventually be improved for a more complex and comprehensive model adding others mechanical, physical or geometrical geo-referenced parameters.

Cone-tip penetration resistances and shear wave velocities have been measured along 16 bore holes spreaded throughout the clay deposits of Mexico City (Figure 2). This information was used as the set of examples inputs (latitude, longitude and depth) → output (qcqc /VsVs). The analysis was carried out in an approximate area of 125 km2km2 of Mexico City downtown. It is important to point out that 20% of these patterns (sample points and complete variables information) are not used in the training stage; they will be presented for testing the generalization capabilities of the closed system components (once the training is stopped).

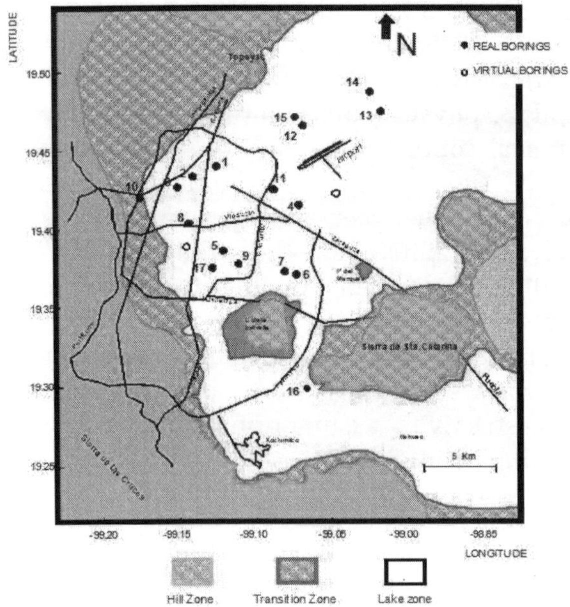

Number	SITE
1	Tlatelolco
2	Alameda
3	Plaza Córdoba
4	Velódromo
5	SCT
6	CAF
7	CDAO
8	CUPJ
9	Eugenia
10	El Águila
11	Línea B
12	Av. 510
13	Calle Urano
14	Plaza Aragón
15	Río Remedios
16	Tláhuac
17	5 de Febrero

FIGURE 2. Mexico City Zonation

In the 3D-neurogenetic analysis, the functions qc={qc(X,Y,Z)}/Vs={Vs(X,Y,Z)} are to be approximated using the procedure outlined below:

1. Generate the database including identification of the site [borings or stations] (X,Y –geographical coordinates, Z –depth, and a CODE –ID number), elevation reference (meters above de sea level, m.a.s.l.), thickness of predetermined structures (layers), and additional information related to geotechnical zoning that could be useful for results interpretation.

2. Use the database to train an initial neural topology whose weights and layers are tuned by an evolutive algorithm (see [46] for details), until the minimum error between calculated and measured values qc=fNN(X,Y,Z)}/Vs={fNN(X,Y,Z)} is achieved (Figure 3a). The generalization capabilities of the optimal 3D neural model are tested presenting real work cases (information from borings not included in the training set) to the net. Figure 3b presents the comparison between the measuredqcqc, VsVsvalues and the NN calculations for

testing cases. Through the neurogenetic results for unseen situations we can conclude that the procedure works extremely well in identifying the general trend in materials resistance (stiffness). The correlation between NN calculations and "real" values is over 0.9.

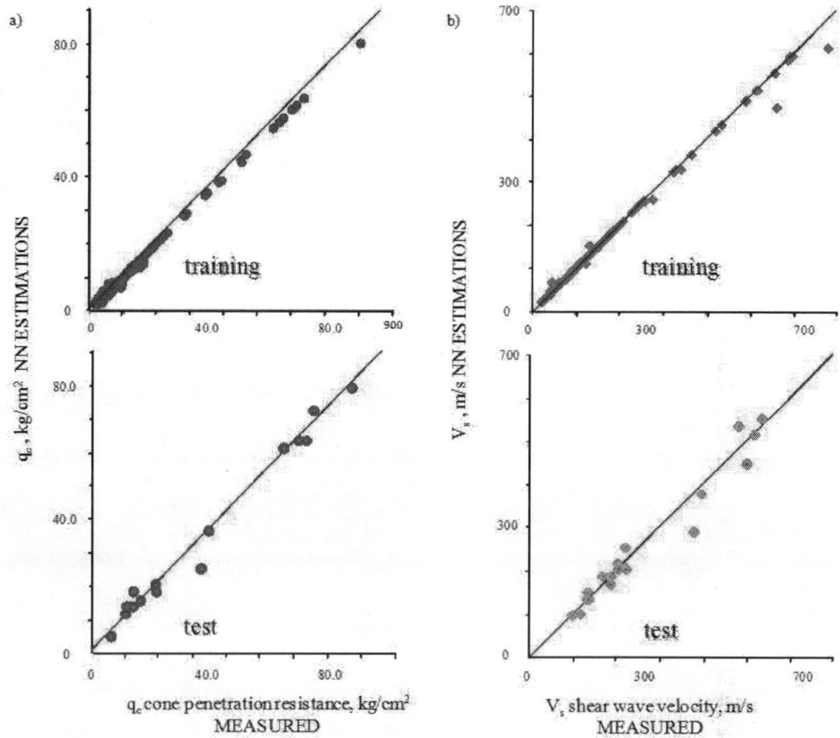

FIGURE 3. Neural estimations of mechanical and dynamic parameters

1. For visual environment requirements a grid is constructed using raw information and neurogenetic estimations for defining the spatial variation of properties (Figure 4). The 3D view of the studied zone represents an easier and more understandable engineering system. The 3D neurogenetic-database also permits to display property-contour lines for specific depths. Using the neurogenetic contour maps, the spatial distribution of the mechanical/dynamic variables can

be visually appreciated. The 3D model is able to reflect the stratigraphical patterns (Figure 5), indicating that the proposed networks are effective in site characterization with remarkable advantages if comparing with geostatistical approximations: it is easier to use, to understand and to develop graphical user interfaces. The confidence and practical advantages of the defined neurogenetic layers is evident. Precision of predictions depends on neighborhood structure, grid size, and variance response, but based on the results we can conclude that despite of the grid cell (size) is not too small the spatial correlation extends beyond the training neighborhood, but the higher confidence is obviously only within.

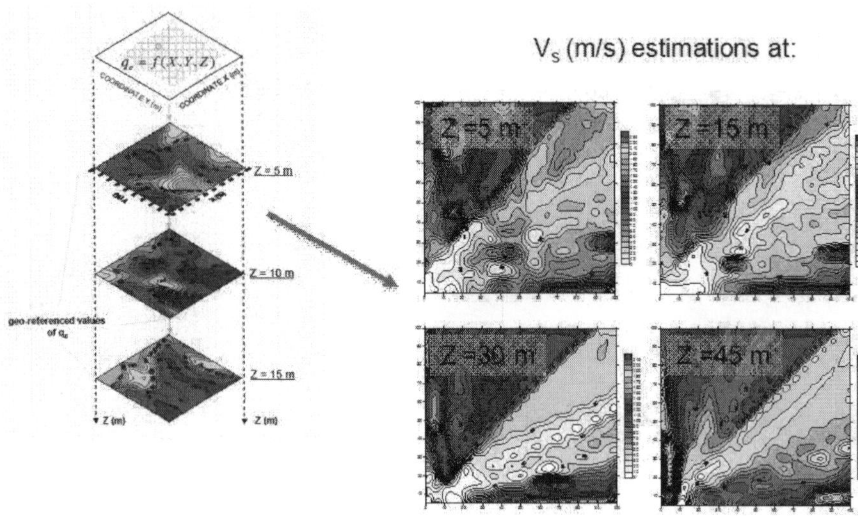

FIGURE 4. Neural response

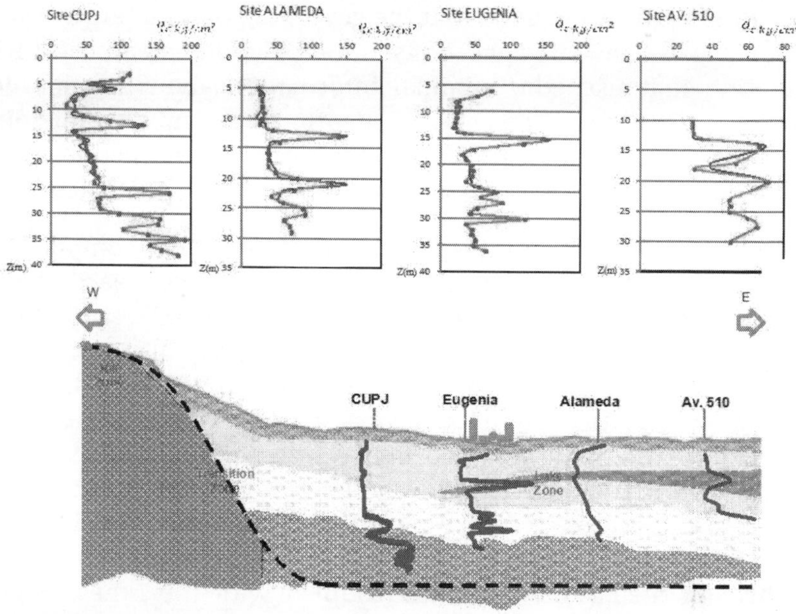

FIGURE 5. Stratigraphy sequence obtained using the 3D Neural estimations

4.2. ATTENUATION LAWS FOR ROCK SITE (OUTCROPPING MOTIONS)

Source, path, and local site response are factors that should be considered in seismic hazard analyses when using attenuation relations. These relations, obtained from statistical regression, are derived from strong motion recordings to define the occurrence of an earthquake with a specific magnitude at a particular distance from the site. Because of the uncertainties inherent in the variables describing the source (e.g. magnitude, epicentral distance, focal depth and fault rupture dimension), the difficulty to define broad categories to classify the site (e.g. rock or soil) and our lack of understanding regarding wave propagation processes and the ray path characteristics from source to site, commonly the predictions from attenuation regression analyses are inaccurate. As an effort to recognize these aspects, multiparametric attenuation relations have been proposed by several researchers [47-53]. However, most of these authors have concluded that the governing parameters are still source, ray path, and site conditions. In this section an empirical NN formulation that uses the minimal information about magnitude, epicentral distance, and focal depth for subduction-zone earthquakes is developed to predict the peak ground acceleration PGA and spectral accelerations SaSa at a rock-like site in Mexico City.

The NN model was training from existing information compiled in the Mexican strong motion database. The NN uses earthquake moment magnitudeMwMw, epicentral distanceEDED, and focal depth FDFD from hundreds of events recorded during Mexican subduction earthquakes (Figure 6) from 1964 to 2007. To test the predicting capabilities of the neuronal model, 186 records were excluded from the data set used in the learning phase. Epicentral distance EDED is considered to be the length from the point where fault-rupture starts to the recording site, and the focal depth FDFD is not declared as mechanism classes, the NN should identify the event type through the FDFD crisp value coupled with the others input parameters [54, 47, 55], The interval of MwMw goes from 3 to 8.1 approximately and the events were recorded at near (a few km) and far field stations (about 690 km). The depth of the zone of energy release ranged from very shallow to about 360 km.

Modeling of the data base has been performed using backpropagation learning algorithm. Horizontal (mutually orthogonal PGAh1 , N-S component, and PGAh2 , E-W component) and vertical components (PGAv) are included as outputs for neural mapping. After trying many topologies, the best horizontal and vertical modules with quite acceptable approximations were the simpler alternatives (BP backpropagation, 2 hidden layers/15 units or nodes each). The neuronal attenuation model for $\{Mw,ED,FD\} \rightarrow \{PGAh1,PGAh2,PGAv\}$ was evaluated by performing testing analyses. The predictive capabilities of the NNs were verified by comparing the estimated PGA's to those induced by the 186 events excluded from the original database (data for training stage). In Figure 7 are compared the computed PGA's during training and testing stages to the measured values. The relative correlation factors (R2≆0.97), obtained in the training phase, indicate that those

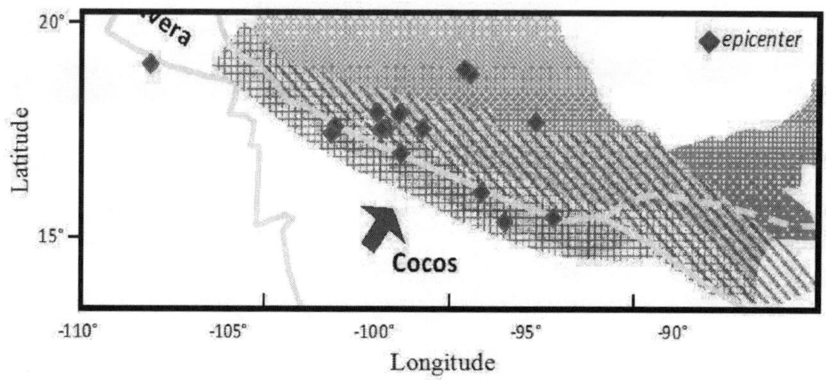

EVENT SUMMARY	
Events	80 from 1964 to 2007
Epicentral coordinates	Latitude from 13.98° to 18.74°
	Longitude from 92.79° to 104.67°
Magnitudes	from 3.9 to 8.1
Focaldepth	from <3 to 360 km
Epicentral distance	from 112 to 690 km

FIGURE 6. Earthquakes characteristics

topologies selected as optimal behave consistently within the full range of intensity, distances and focal depths depicted by the patterns. Once the networks converge to the selected stop criterion, learning is finished and each of these black-boxes become a nonlinear multidimensional functional. Following this procedure 20 NN are trained to evaluate de SaSa at different response spectra periods (from T= 0.1 s to T= 5.0 s with DT=0.25 s). Forecasting of the spectral components is reliable enough for practical applications.

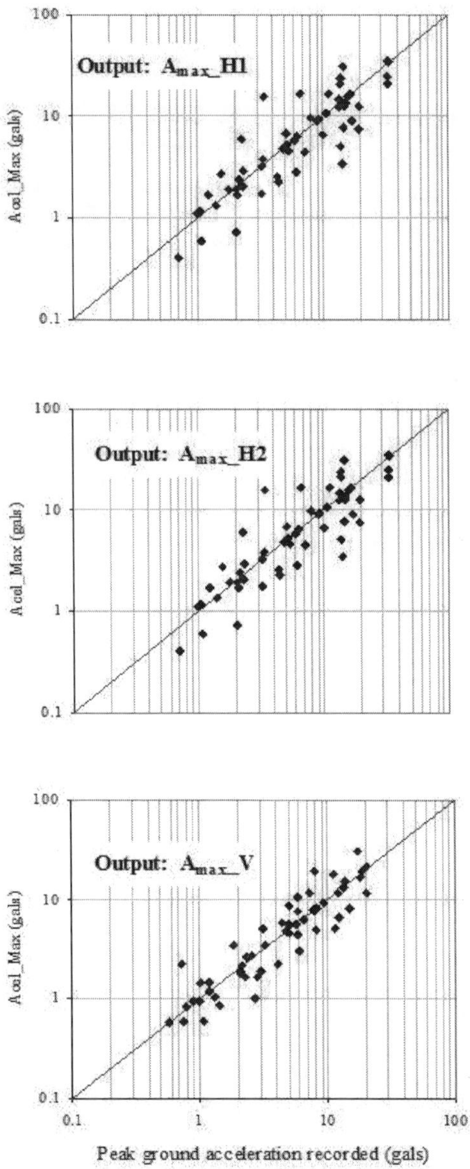

FIGURE 7. Some examples of measured and NN-estimated PGA values

In Figure 8 two case histories correspond to large and medium size events are shown, the estimated values obtained for these events using the relationships proposed by Gómez, Ordaz &Tena [56], Youngs et al. [47], Atkinson and Boore [55] –proposed for rock sites– and Crouse et al. [51] –proposed for stiff soil sites– and the predictions obtained with the PGAh1–h2PGAh1-h2modules are shown. It can be seen that the estimation obtained with Gómez, Ordaz y Tena [56] seems to underestimate the response for the large magnitude event. However, for the lower magnitude event follows closely both the measured responses and NN predictions. Youngs et al. [47] attenuation relationship follows closely the overall trend but tends to fall sharply for long epicentral distances.

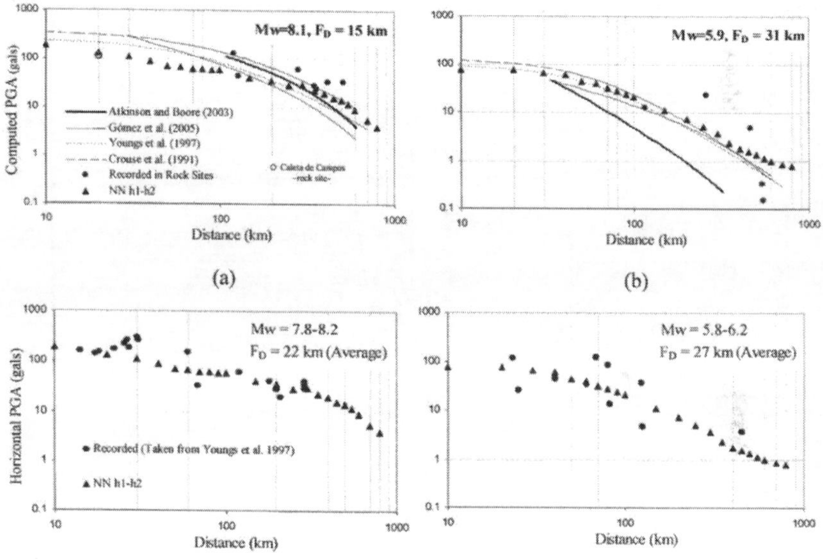

FIGURE 8. Attenuation laws comparisons

Furthermore, it should be stressed the fact that, as can be seen in Figure 9 the neural attenuation model is capable to follow the general behavior of the measure data expressed as spectra while the traditional functional approaches are not able to reproduce. A neural sensitivity study for the input variables was conducted for the neuronal modules. The results are strictly valid only for the data base utilized, nevertheless, after several sensitivity analyses conducted changing the database composition, it was found that the following trend prevails; theMwMw would be the most relevant parameter then would follow EDED coupled with FDFD. However, for near site events the epicentral distance could become as relevant as the magnitude, particularly, for the vertical component and for minor earthquakes (M low) the FDFD becomes very transcendental.

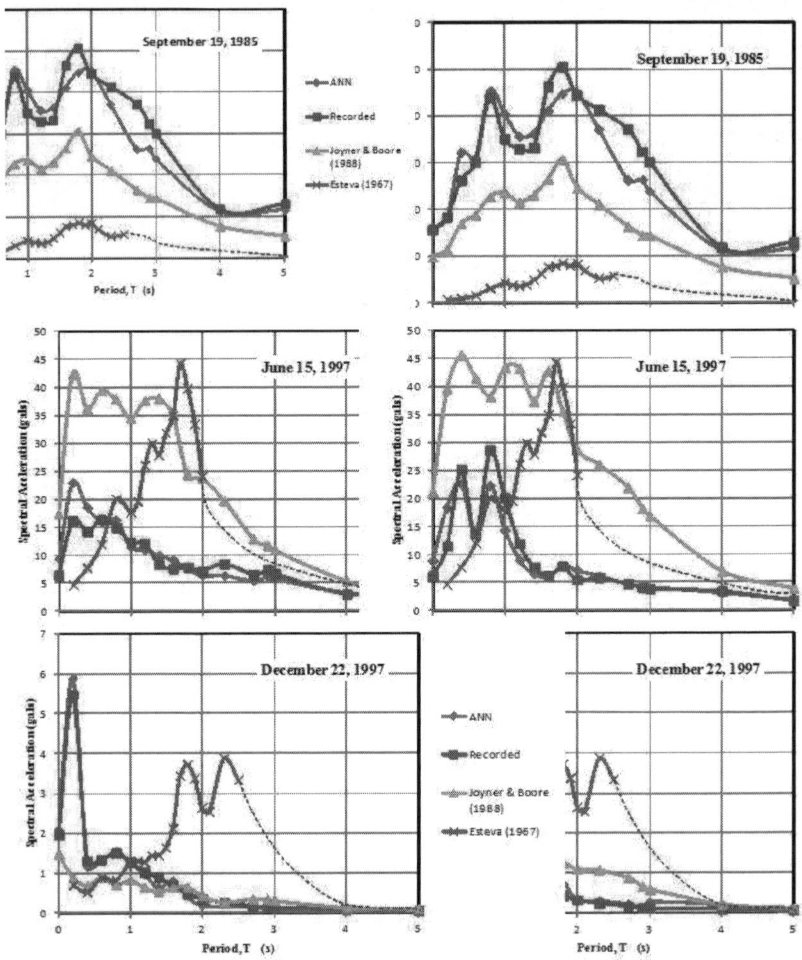

FIGURE 9. Response Spectra: NN-calculated vs traditional functions

Through{Mw,ED,FD}→{PGAhi,Sa}{Mw,ED,FD}→{PGAhi,Sa} mapping, this neuronal approach offers the flexibility to fit arbitrarily complex trends in magnitude and distance dependence and to recognize and select among the tradeoffs that are present in fitting the observed parameters within the range of magnitudes and distances present in data. This approach seems to be a promising alternative to describe earthquake phenomena despite of the limited observations and qualitative knowledge of the recording stations geotechnical site conditions, which leads to a reasoning of a partially defined behavior.

4.3. GENERATION OF ARTIFICIAL TIME SERIES: ACCELEROGRAMS APPLICATION

For nonlinear seismic response analysis, where the superposition techniques do not apply, earthquake acceleration time histories are required as inputs. Virtually all seismic design codes and guidelines require scaling of selected ground motion time histories so that they match or exceed the controlling design spectrum within a period range of interest. Considerable variability in the characteristics of the recorded strong-motions under similar conditions may still require a characterization of future shaking in terms of an ensemble of accelerograms rather than in terms of just one or two "typical" records. This situation has thus created a need for the generation of synthetic (artificial) strong-motion time histories that simulate realistic ground motions from different points of views and/or with different degrees of sophistication. To provide the ground motions for analysis and design, various methods have been developed: i) frequency-domain methods where the frequency content of recorded signals is manipulated [57-60] and ii) time-domain methods where the recorded ground motions amplitude is controlled [61, 62]. Regardless of the method, first, one or more time histories are selected subjectively, and then scaling mechanisms for spectrum matching are applied. This is a trial and error procedure that leads artificial signals very far from real-earthquake time series.

In this investigation a Genetic Generator of Signals is presented. This genetic generator is a tool for finding the coefficients of a pre-specified functional form, which fit a given sampling of values of the dependent variable associated with particular given values of the independent variable(s). When the genetic generator is applied to synthetic accelerograms construction, the proposed tool is capable of i) searching, under specific soil and seismic conditions (within thousands of earthquake records) and recommending a desired subset that better match a target design spectrum, and ii) through processes that mimic mating, natural selection, and mutation, producing new generations of accelerograms until an optimum individual is obtained. The procedure is fast and reliable and results in time series that match any type of target spectrum with minimal tampering and deviation from recorded earthquakes characteristics.

The objective of the genetic generator, when applied to synthetic earthquakes construction, is to produce compatible artificial signals with specific design spectra. In this model specific seismic (fault rupture, magnitude, distance, focal depth) and site characteristics (soil/ rock) are the first set of inputs. They are included to take into consideration that a typical strong motion record consists of a variety of waves whose contribution depends on the earthquake source mechanism (wave path) and its particular characteristics are influenced by the distance between the source and the site, some measure of the size of the earthquake, and the surrounding geology and site conditions; and that the design spectra can be an envelope or integration of many expected ground motions that are possible to occur in certain period of time, or the result of a formulation that involves earthquake magnitude, distance and soil conditions. The second set of inputs consist of the target spectrum, the period range for the matching, lower- and upper-bound acceptable values for scaling signal shape, and a collection of GAs parameters (a population size, number of generations, crossover ratio, and mutation ratio). The output is the more success individual with a chromosome array generated from "real" accelerograms parents (a set of).

The algorithm (see Figure 10) is started with a set of solutions (each solution is called a chromosome). A solution is composed of thousands of components or genes (accelerations recorded at the time), each one encoding a particular trait. The initial solutions (original population) are selected based on the seismic parameters at a site (defined previously by the user): fault mechanism, moment magnitude, epicentral distance, focal depth, geotechnical and geological site classification, depth of sediments. If the user does not have a priori seismic/site knowledge, the genetic generator could select the initial population randomly (Figure 11). Once the model has found the seed-accelerogram(s) or chromosome(s), the space of all feasible solutions can be called accelerograms space (state space). Each point in this search space represents one feasible solution and can be "marked" by its value or fitness for the problem. The looking for a solution is then equal to a looking for some extreme (minimum or maximum) in the space.

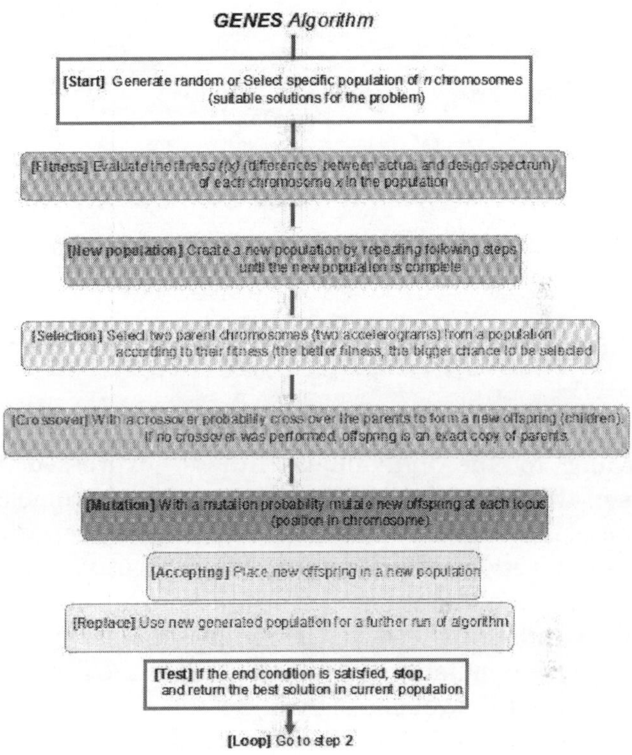

FIGURE 10. Genetic Generator: flow diagram

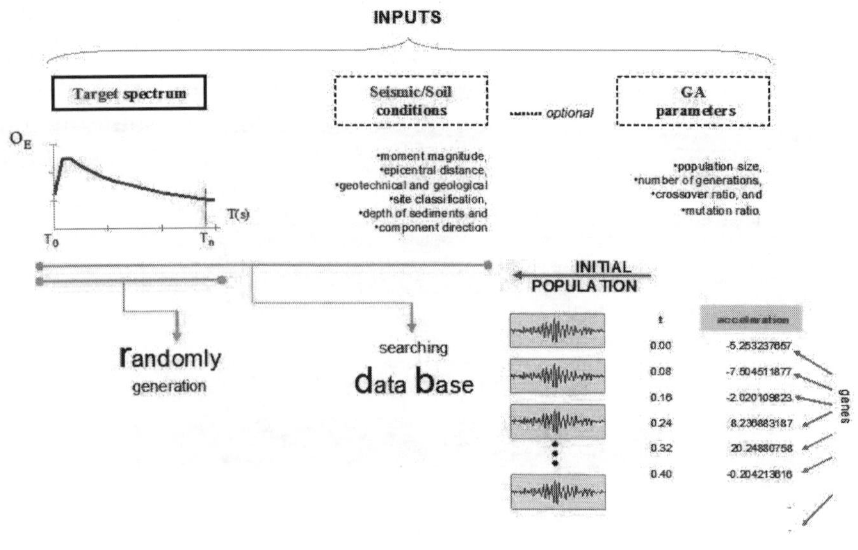

FIGURE 11. Genetic Generator: working phase diagram

According to the individuals' fitness, expressed by difference between the target design spectrum and the chromosome response spectrum, the problem is formulated as the minimization of the error function, Z, between the actual and the target spectrum in a certain period range. Solutions with highest fitness are selected to form new solutions (offspring). During reproduction, the recombination (or crossover) and mutation permits to change the genes (accelerations) from parents (earthquake signals) in some way that the whole new chromosome (synthetic signal) contains the older organisms attributes that assure success. This is repeated until some user's condition (for example number of populations or improvement of the best solution) is satisfied (Figure 12).

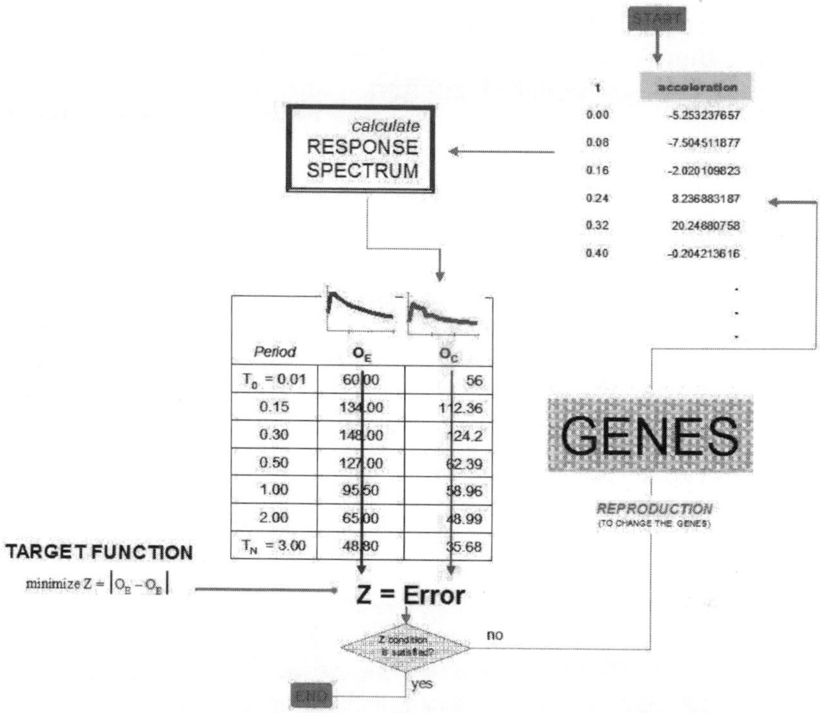

FIGURE 12. Iteration process of the Genetic Generator

One of the genetic advantages is the possibility of modifying on line the image of the expected earthquake. While the genetic model is running the user interface shows the chromosome per epoch and its response spectra in the same window, if the duration time, the highest intensities interval or the ØtØtare not convenient for the user's interests, these values can be modified without retraining or a change on model' structure.

In Figure 13 are shown three examples of signals recovered following this methodology. The examples illustrate the application of the genetic methodology to select any number of records to match a given target spectrum (only the more successful individuals for each target are shown in the figure). It can be noticed the stability of the genetic algorithm in adapting itself to smooth, code or scarped spectrum shapes. The procedure is fast and reliable as results in records match the target spectrum with minimal deviation. The genetic procedure has been applied successfully to generate synthetic ground motions having different amplitudes, duration and combinations of moment magnitude and epicentral distance. Although the variations in the target spectra, the genetic signals maintain the nonlinear and nonstationary characteristics of real earthquakes. It is still under development an additional toolbox that will permit to use advanced signal analysis instruments because, as it has been demonstrated [63] [64], studying nonstationary signals through Fourier or response spectra is not convenient for all applications.

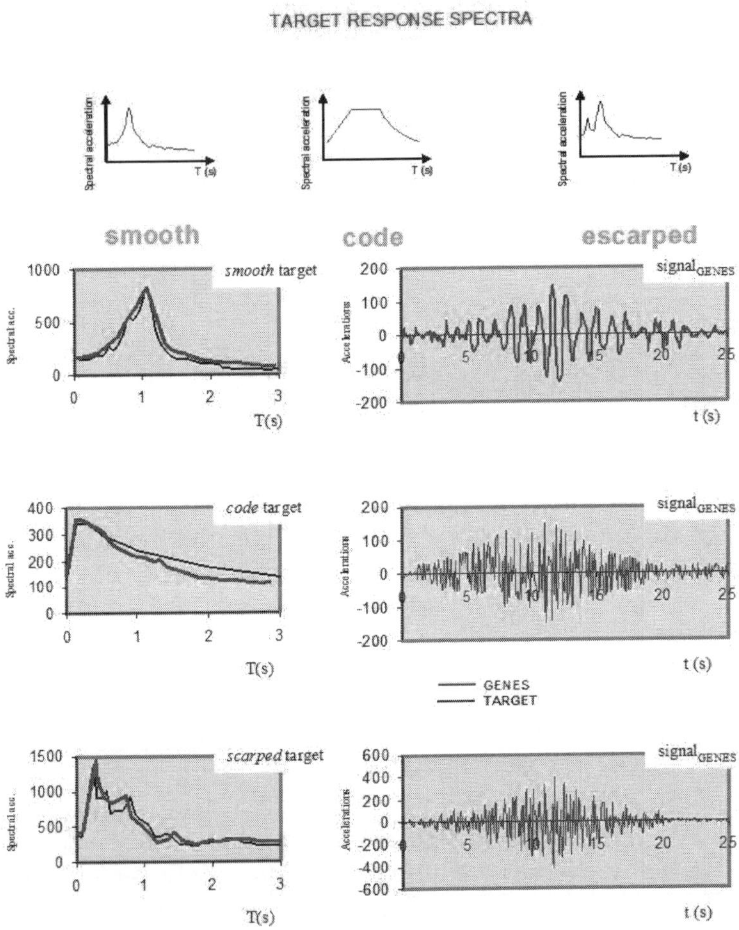

FIGURE 13. Some Generator results: accelerograms application

4.4. EFFECTS OF LOCAL SITE CONDITIONS ON GROUND MOTIONS

Geotechnical and structural engineers must take into account two fundamental characteristics of earthquake shaking: 1) how ground shaking propagates through the Earth, especially near the surface (site effects), and 2) how buildings respond to this ground motion. Because neither characteristic is completely understood, the seismic phenomenon is still a challenging research area.

Site effects play a very important role in forecasting seismic ground responses because they may strongly amplify (or deamplify) seismic motions at the last moment just before reaching the surface of the ground or the basement of man-made structures. For much of the history of seismological research, site effects have received much less attention than they should, with the exception of Japan, where they have been well recognized through pioneering work by Sezawa and Ishimoto as early as the 1930's [65]. The situation was drastically changed by the catastrophic disaster in Mexico City during the Michoacan, Mexico earthquake of 1985, in which strong amplification due to extremely soft clay layers caused many high-rise buildings to collapse despite their long distance from the source. The cause of the astounding intensity and long duration of shaking during this earthquake is not well resolved yet even though considerable research has been conducted since then, however, there is no room for doubt that the primary cause of the large amplitude of strong motions in the soft soil (lakebed) zone relative to those in the hill zone is a site effect of these soft layers.

The traditional data-analysis methods to study site effects are all based on linear and stationary assumptions. Unfortunately, in most soil systems, natural or manmade ones, the data are most likely to be both nonlinear and nonstationary. Discrepancies between calculated responses (using code site amplification factors) and recent strong motion evidence point out serious inaccuracies may be committed when analyzing amplification phenomena. The problem might be due partly because of the lack of understanding regarding the fundamental causes in soil response but also a consequence of the distorted soil amplification quantification and the incomplete characterization of nonlinearity-induced nonstationary features exposed in motion recordings [66]. The objective of this investigation is to illustrate a manner in which site effects can be dealt with for the case of Mexico City soils, making use of response spectra calculated from the motions recorded at different sites during extreme and minor events (see Figure 6). The variations in the spectral shapes, related to local site conditions, are used to feed a multilayer neural network that represent a very advantageous nonlinear-amplification relation. The database is composed by registered information earthquakes affecting Mexico City originated by different source mechanisms.

The most damaging shocks, however, are associated to the subduction of the Cocos Plate into the Continental Plate, off the Mexican Pacific Coast. Even though epicentral distances are rather large, these earthquakes have recurrently damaged structures and produced severe losses in Mexico City. The singular geotechnical environment that prevails in Mexico City is the one most important factor to be accounted for in explaining the huge amplification of seismic movements [67-70]. The soils in Mexico City were formed by the deposition into a lacustrine basin of air and water transported materials. From the view point of geotechnical engineering, the relevant strata extend down to depths of 50 m to 80 m, approximately. The superficial layers formed the bed of a lake system that has been subjected to dessication for the last 350 years. Three types of soils may be broadly distinguished: in Zone I, firm soils and rock-like materials prevail; in Zone III, very soft clay formations with large amounts of microorganisms interbedded by thin seams of silty sands, fly ash and volcanic glass are found; and in Zone II, which is a transition between zones I and III, sequences of clay layers an coarse material strata are present (Figure 14).

Due to space limitations, reference is made only to two seismic events: the June 15, 1999 and the October 9, 1995. This module was developed based in a previous study (see section 4.2 of this Chapter) where the effect of the parametersEDED, FDFD and MwMw on the ground motion attenuation from epicentre to the site, were found to be the most significant [71]. The recent disaster experience showed that the imprecision that is inherent to most variables

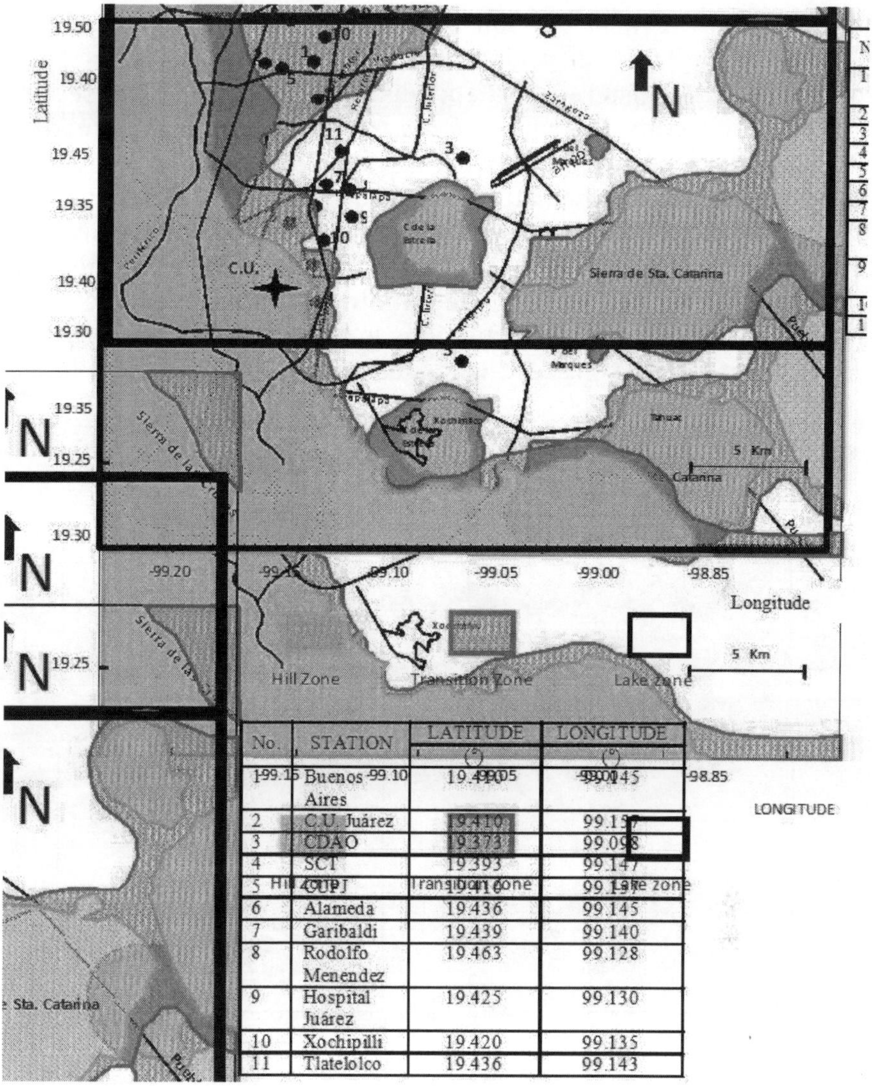

No.	STATION	LATITUDE (°)	LONGITUDE (°)
1	Buenos Aires	[illegible]	99.145
2	C.U. Juárez	19.410	99.15[illegible]
3	CDAO	19.373	99.09[illegible]
4	SCT	19.393	99.147
5	CUPJ	[illegible]	[illegible]
6	Alameda	19.436	99.145
7	Garibaldi	19.439	99.140
8	Rodolfo Menendez	19.463	99.128
9	Hospital Juárez	19.425	99.130
10	Xochipilli	19.420	99.135
11	Tlatelolco	19.436	99.143

FIGURE 14. Accelerographic stations used in this study

measurements or estimations makes crucial the consideration of subjectivity to evaluate and to derive numerical conclusions according to the phenomena behavior. The neuronal training process starts with the training of four input variables booked:EDED, FDFD and MwMw. The output linguistic variables are PGAh1PGAh1 (peak ground acceleration horizontal, component 1) and ,PGAh2,PGAh2 (peak ground acceleration, horizontal component 2) registered in a rock-like site in Zone I .The second training process is linked *feed-forward* with the previous module (PGA for rock-like site) and the new seismic inputs are Seismogenic Zone and PGA_{rock} and the Latitude and Longitude coordinates are the geo-referenced position needed to draw the deposition variation into the basin. This neuro-training runs one step after the first training phase and until the minimum difference between theSaSa and the neuronal calculations is attained. In Figure 15 some results from training and testing modes are shown.

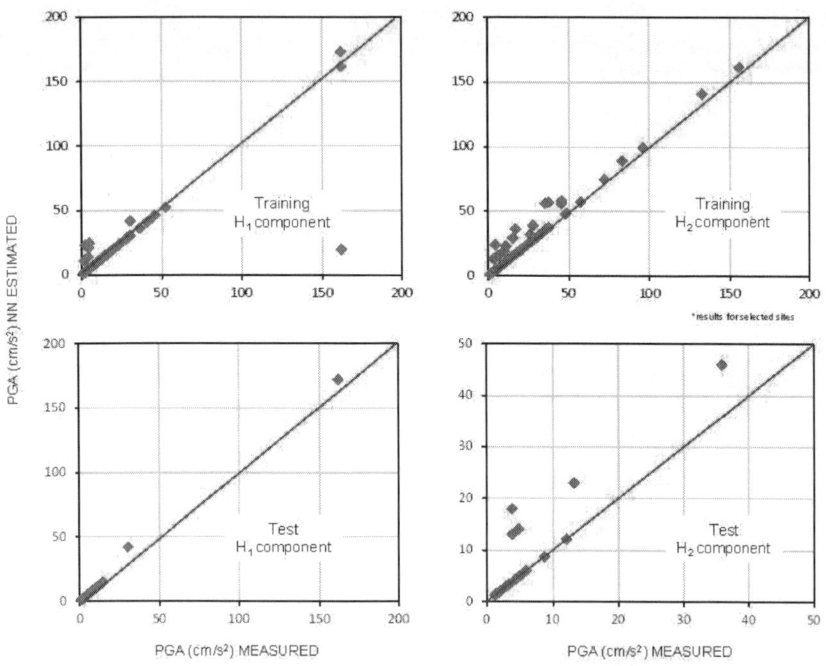

FIGURE 15. Neural estimations for PGA in Lake Zone sites

This second NN represents the geo-referenced amplification ratio that take into consideration the topographical, geotechnical and geographical conditions, implicit in the recorded accelerograms.The results of these two NNs are summarized in Figure 16. These graphs show the predicting capabilities of the neural system comparing the measured values with those obtained in neural-working phase. It can be observed a good correspondence throughout the full distance and magnitude range for the seismogenic zones considered in this study for the whole studied area (Lake Zone).

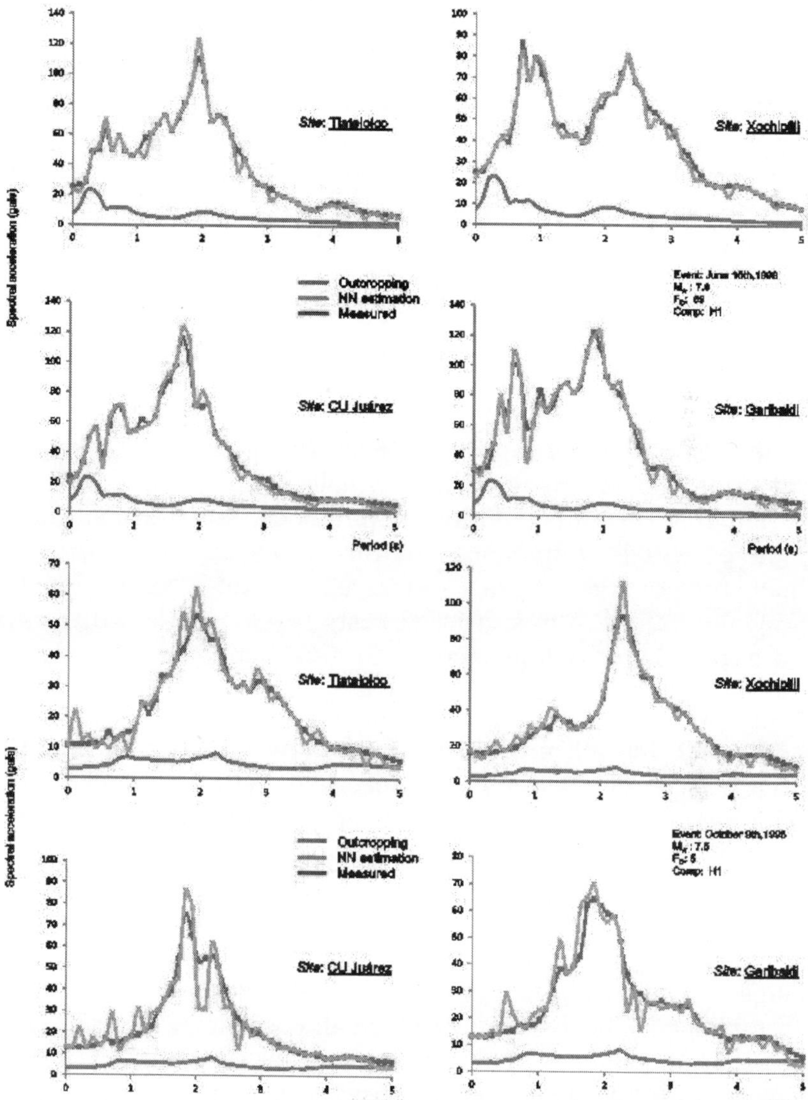

FIGURE 16. Spectral accelerations in some Lake-Zone sites: measured vs NN

4.5. LIQUEFACTION PHENOMENA: POTENTIAL ASSESSMENT AND LATERAL DISPLACEMENTS ESTIMATION

Over the past forty years, scientists have conducted extensive research and have proposed many methods to predict the occurrence of liquefaction. In the beginning, undrained cyclic loading laboratory tests had been used to evaluate the liquefaction potential of a soil [72] but due to difficulties in obtaining undisturbed samples of loose sandy soils, many researchers have preferred to use *in situ* tests [73]. In a semi-empirical approach the theoretical considerations and experimental findings provides the ability to make sense out of the field observations, tying them together, and thereby having more confidence in the validity of the approach as it is used to interpolate or extrapolate to areas with insufficient field data to constrain a purely empirical solution. Empirical field-based procedures for determining liquefaction potential have two critical constituents: i) the analytical framework to organize past experiences, and ii) an appropriate *in situ* index to represent soil liquefaction characteristics. The original simplified procedure [74] for estimating earthquake-induced cyclic shear stresses continues to be an essential component of the analysis framework. The refinements to the various elements of this context include improvements in the in-situ index tests (e.g., SPT, CPT, BPT,VsVs), and the compilation of liquefaction/no-liquefaction cases.

The objective of the present study is to produce an empirical machine learning ML method for evaluating liquefaction potential. ML is a scientific discipline concerned with the design and development of algorithms that allow computers to evolve behaviours based on empirical data, such as from sensor data or databases. Data can be seen as examples that illustrate relations between observed variables. A major focus of ML research is to automatically learn to recognize complex patterns and make intelligent decisions based on data; the difficulty lies in the fact that the set of all possible behaviours given all possible inputs is too large to be covered by the set of observed examples (training data). Hence the learner must generalize from the given examples, so as to be able to produce a useful output in new cases. In the following two ML tools, Neural Networks NN and Classification Trees CTs, are used to evaluate liquefaction potential and to find out the liquefaction control parameters, including earthquake and soil conditions. For each of these parameters, the emphasis has been on developing relations that capture the essential physics while being as simplified as possible. The proposed cognitive environment permits an improved definition of i) seismic loading or cyclic stress ratio CSR, and ii) the*resistance* of the soil to triggering of liquefaction or cyclic resistance ratio CRR.

The factor of safety FS against the initiation of liquefaction of a soil under a given seismic loading is commonly described as the ratio of cyclic resistance ratio (CRR), which is a measure of liquefaction resistance, over cyclic stress ratio (CSR), which is a representation of seismic loading that causes liquefaction, symbolically, FS=CRR/CSRFS=CRR/CSR. The reader is referred to Seed and Idriss [74], Youd et al.[75], and Idriss and Boulanger [76] for a historical perspective of this approach. The term CSR CSR=f(0.65,σvo,amax,σ'vo,rd,MSF)CSR=f0.65,σvo,amax,σvo',rd,MSFis function of the vertical total stress of the soil σvoσvo at the depth considered, the vertical effective stress σ'voσvo', the peak horizontal ground surface acceleration amaxamax, a depth-dependent shear stress reduction factor rdrd (dimensionless), a magnitude scaling factor MSFMSF(dimensionless). For CRR, different in situ-resistance measurements and overburden correction factors are included in its determination; both terms operate depending of the geotechnical conditions. Details about the theory behind this topic in Idriss and Boulanger, [76] and Youd et al. [75].

Many correction/adjustment factors have been included in the conventional analytical frameworks to organize and to interpret the historical data. The correction factors improve the consistency between the geotechnical/seismological parameters and the observed liquefaction behavior, but they are a consequence of a constrained analysis space: a 2D plot [CSR *vs.* CRR] where regression formulas (simple equations) intend to relate complicated nonlinear/multidimensional information. In this investigation the ML methods are applied to discover unknown, valid patterns and relationships between geotechnical, seismological and engineering descriptions using the relevant available information of liquefaction phenomena (expressed as empirical prior knowledge and/or input-output data). These ML techniques "work" and "produce" accurate predictions based on few logical conditions and they are not restricted for the mathematical/analytical environment. The ML techniques establish a*natural* connection between experimental and theoretical findings.

Following the format of the simplified method pioneered by Seed and Idriss [74], in this investigation a nonlinear and adaptative *limit state* (a fuzzy-boundary that separates liquefied cases from nonliquefied cases) is proposed (Figure 17). The database used in the present study was constructed using the information included in Table 2 and it was compiled by Agdha et al., [77], Juang et al., [78], Juang [79], Baziar, [80] and Chern and Lee [81]. The cases are derived from cone penetration tests CPT, and shear wave velocities VsVs measurements and different world seismic conditions (U.S., China, Taiwan, Romania, Canada and Japan). The soils types ranges from clean sand and silty sand to silt mixtures (sandy and clayey silt). Diverse geological and geomorphological characteristics are included. The reader is referred to the citations in Table 2 for details.

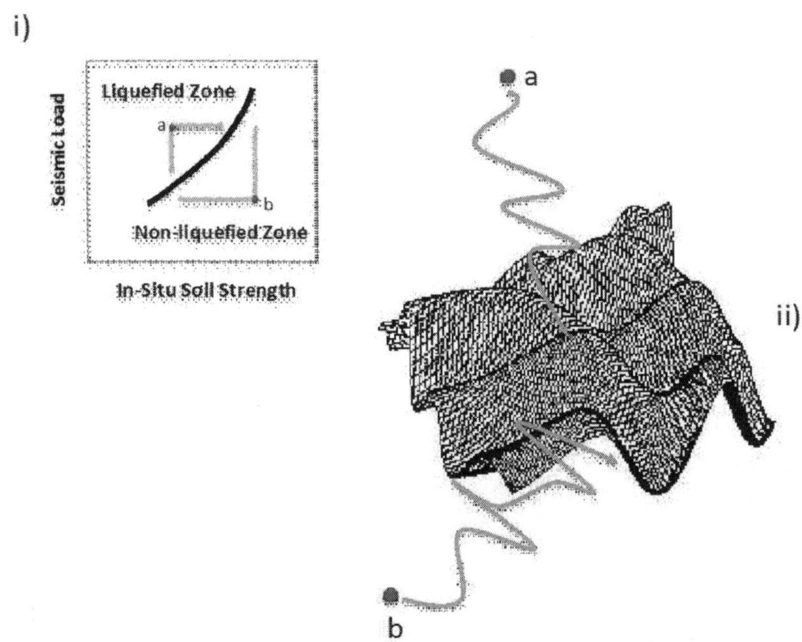

FIGURE 17. An schematic view of the nonlinear-liquefaction boundary

The ML formulation uses Geotechnical (qcqc, VsVs, Unit weight, Soil Type, Total vertical stresses, Effective vertical stresses, Geometrical (Layer thickness, Water Level Depth, Top Layer Depth) andSeismological (Magnitude, PGA) input parameters and the output variable is "Liquefaction?" and it can take the values "YES/NO" (Figure 17). Once the NN is trained the number of cases that was correctly evaluated was 100% and applied to "unseen" cases (separated for testing) less than 10% of these examples were not fitted. The CT has a minor efficiency during the training showing 85% of cases correctly predicted, but when the CT runs on the unseen patterns its capability is not diminished and it asserts the same proportion. From these findings it is concluded that the neuro system is capable of predicting the in situ measurements with a high degree of accuracy but if improvement of knowledge is necessary or there are missed, vague even contradictory values in the analyzed case, the CT is a better option.

TABLE 2. Database for liquefaction analysis

Set	Input Parameters	Number .of Patterns	Ref.
A	Z, Z_{NAF}; H, Soil Class, Geomorphological units, Geological units, Site amplification, a_{max}	56	Fatemi-Agdha et al., 1988
B	Z, q_c, F_s, σ_0, σ_0', a_{max}, M	21	Juang et al., 1999
C	Z, q_c, F_s, σ_0, σ_0', a_{max}, M	242	Juang, 2003
D	D_{50}, a_{max}, σ_0', σ_0, M, F_s, q_c, SPT, Z	170	Baziar, 2003
E	M, σ_0, σ_0', q_c, a_{max}	466	Chern and Lee, 2009
F	Z_{NAF}, Z, H, σ_0, σ_0', Soil Class, V_s	80	Andrus and Stokoe, 1997; 2000
	Total:	**1035**	

Figure 18 shows the pruned liquefaction trees (two, one runs using qcqc values and the other through theVsVs measurements) with YES/NO as terminal nodes. In the Figure 19, some examples of tree reading are presented. The trees incorporate soil type dependence through the resistance values (qcqc, andVsVs) and fine content, and it is not necessary to label the material as "sand" or "silt". The most general geometrical branches that split the behaviors are the Water table depth and the Layer thickness but only when the soil description is based onVsVs, whenqcqc, serves as rigidity parameter this geometrical inputs are not explicit exploited. This finding can be related to the nature of the measurement: the cone penetration value contains the effect of the saturated material while the shear wave velocities need the inclusion of this situation explicitly. Without potentially confusing regression strategies, the liquefaction trees results can be seen as an indication of how effectively the ML model maps the assigned predictor variables to the response parameter. Using data from all regions and wide parameters ranges, the prediction capabilities of the neural network and classification trees are superior to many other approximations used in common practice, but the most important remark is the generation of meaningful clues about the reliability of physical parameters, measurement and calculation process and practice recommendations.

The intricacy and nonlinearity of the phenomena, an inconsistent and contradictory database, and many subjective interpretations about the observed behavior, make SC an attractive alternative for estimation of liquefaction induced lateral spread. NEFLAS [82], NEuroFuzzy estimation of liquefaction induced LAteral Spread, profits from fuzzy and neural paradigms through an architecture that uses a fuzzy system to represent knowledge in an interpretable manner and proceeds from the learning ability of a neural network to optimize its parameters. This blending can constitute an interpretable model that is capable of learning the problem-specific prior knowledge.

NEFLAS is based on the Takagi-Sugeno model structure and it was constructed according the information compiled by Bartlett and Youd [83] and extended later by Youd et al. [84]. The output considered in NEFLAS is the horizontal displacements due to liquefaction, dependent of moment magnitude, the PGA, the nearest distance from the source in kilometers; the free face ratio, the gradient of the surface topography or the slope of the liquefied layer base, the cumulative thickness of saturated cohesionless sediments with number of blows (modified by overburden and energy delivered to the standard penetration probe, in this case 60%) , the average of fines content, and the mean grain size.

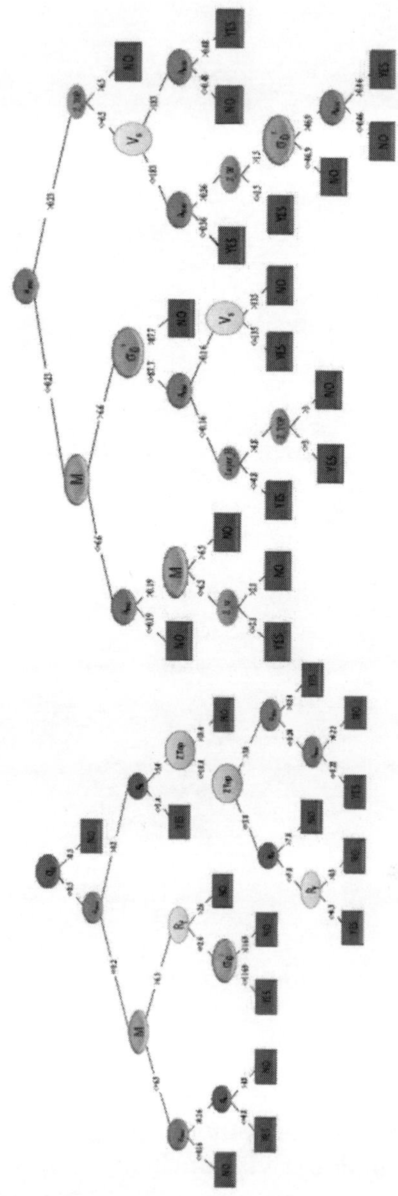

FIGURE 18. Classification tree for liquefaction potential assessment

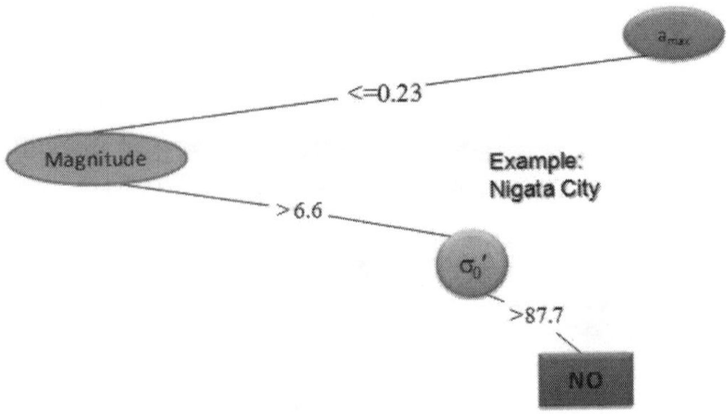

Magnitude (Mw)	Depth (m)	Z_w (m)	Z_TOP (m)	Layer_H (m)	σ'v (kPa)	R_f (%)	Type of Soil	qc (MPA)	Vs (m/s)	amax (g)	Liquefied
7.50	-	5.00	5.00	2.50	97.70	-	silt	-	163.00	0.16	no

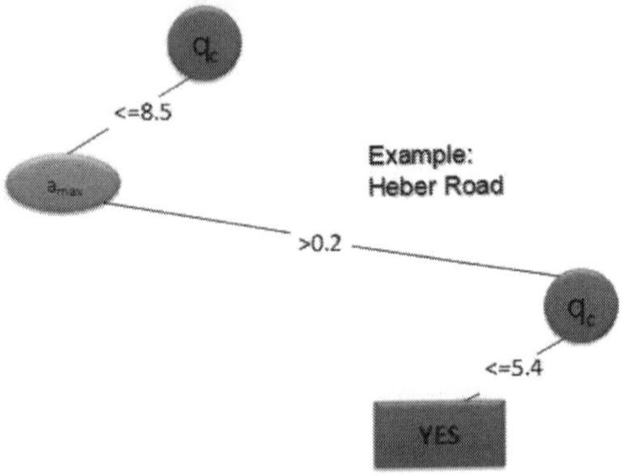

Magnitude (Mw)	Depth (m)	Z_w (m)	Z_TOP (m)	Layer_H (m)	σ'v (kPa)	R_f (%)	Type of Soil	qc (MPA)	Vs (m/s)	amax (g)	Liquefied
6.60	4.00	-	-	-	56.00	2.80	-	2.0	-	0.8	yes

FIGURE 19. CT classification examples

One of the most important NEFLAS advantages is its capability of dealing with the imprecision, inherent in geoseismic engineering, to evaluate concepts and derive conclusions. It is well known that engineers use words to classify qualities ("strong earthquake", "poor graduated soil" or "soft clay" for example), to predict and to validate "first principle" theories, to enumerate phenomena, to suggest new hypothesis and to point the limits of knowledge. NEFLAS mimics this method. See the technical quantity "magnitude" (earthquake input) depicted in Figure 20. The degree to which a crisp magnitude belongs to LOW, MEDIUM or HIGH linguistic label is called the degree of membership. Based on the figure, the expression, "the magnitude is LOW" would be true to the degree of 0.5 for aMwMwof 5.7. Here, the degree of membership in a set becomes the degree of truth of a statement.

On the other hand, the human logic in engineering solutions generates sets of behavior rules defined for particular cases (parametric conditions) and supported on numerical analysis. In the neurofuzzy methods the human concepts are re-defined through a flexible computational process (training) putting (empirical or analytical) knowledge into simple "if-then" relations (Figure 20). The fuzzy system uses 1) variables composing the antecedents (premises) of implications; 2) membership functions of the fuzzy sets in the premises, and 3) parameters in consequences for finding simpler solutions with less design time.

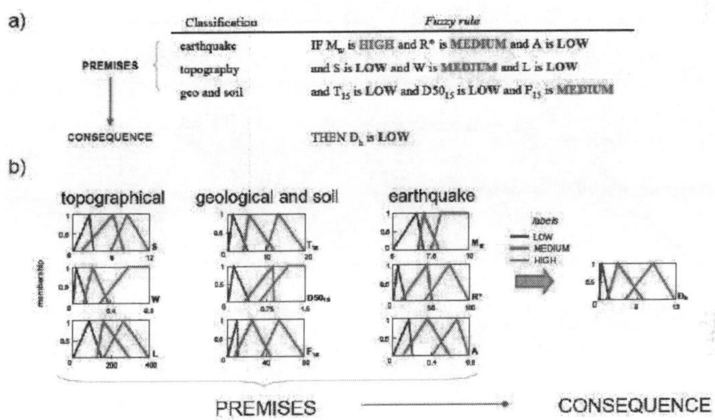

FIGURE 20. Neurofuzzy estimation of lateral spread

NEFLAS considers the character of the earthquake, topographical, regional and geological components that influence lateral spreading and works through three modules: Reg-NEFLAS, appropriate for predicting horizontal displacements in geographic regions where seismic hazard surveys have been identified; Site- NEFLAS, proper for predictions of horizontal displacements for site-specific studies with minimal data on geotechnical conditions and Geotech-NEFLAS allows more refined predictions of horizontal displacements when additional data is available from geotechnical soil borings. The NEFLAS execution on cases not included in the database (Figure 21.b and Figure 21.c) and its higher values of correlation when they are compared with evaluations obtained from empirical procedures permit to assert that NEFLAS is a powerful tool, capable of predicting lateral spreads with high degree of confidence.

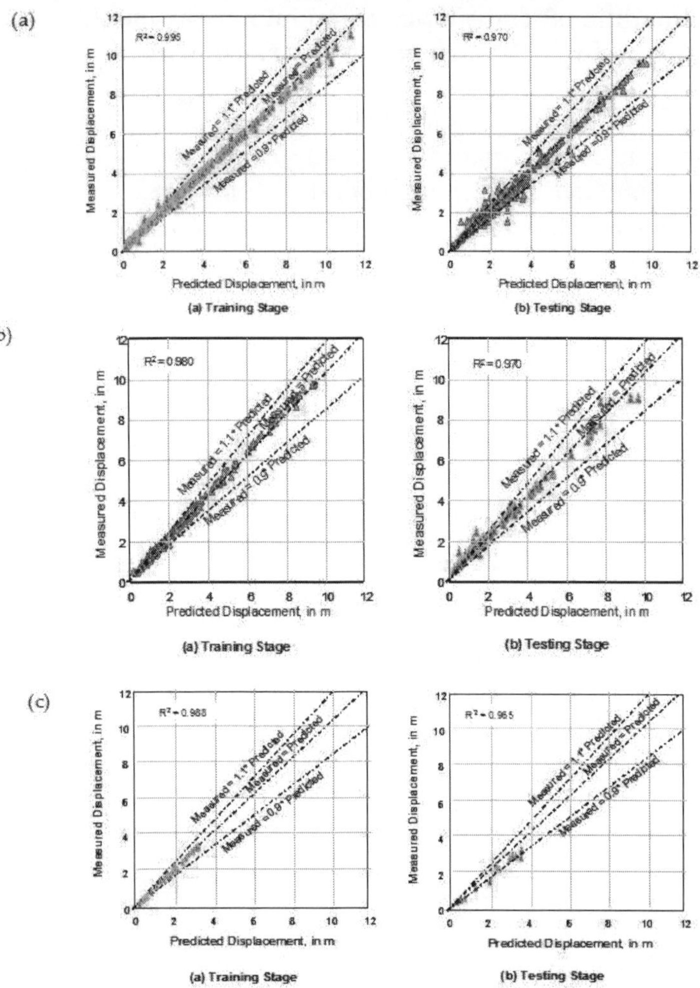

FIGURE 21. NN estimations vs measured displacements for a) the whole data set, b)Niigata Japan, c) San Francisco USA cases

5. CONCLUSIONS

Based on the results of the studies discussed in this paper, it is evident that cognitive techniques perform better than, or as well as, the conventional methods used for modeling complex and not well understood geotechnical earthquake problems. Cognitive tools are having an impact on many geotechnical and seismological operations, from predictive modeling to diagnosis and control.

The hybrid *soft* systems leverage the tolerance for imprecision, uncertainty, and incompleteness, which is intrinsic to the problems to be solved, and generate tractable, low-cost, robust solutions to such problems. The synergy derived from these hybrid systems stems from the relative ease with which we can translate problem domain knowledge into initial model structures whose parameters are further tuned by local or global search methods. This is a form of methods that do not try to solve the same problem in parallel but they do it in a mutually complementary fashion. The push for low-cost solutions combined with the need for intelligent tools will result in the deployment of hybrid systems that efficiently integrate reasoning and search techniques.

Traditional earthquake geotechnical modeling, as physically-based (or knowledge-driven) can be improved using soft technologies because the underlying systems will be explained also based on data (CC data-driven models). Through the applications depicted here it is sustained that cognitive tools are able to make abstractions and generalizations of the process and can play a complementary role to physically-based models.

REFERENCES

1. W.D.L.Finn State of the art of geotechnical earthquake engineering practice Soil Dynamics and Earthquake Engineering 20 (2000Elsevier

2. N. A. R. R. Abrahamson, A. Youngs, algorithm. stable, regression. for, using. analysis, random. the, model. effects, Bull Seismol Soc Am 1992

3. P. G. Somerville, T. Sato, Correlation of rise time with the style-offaulting factor in strong ground motions. Seismol Res Lett 1998pp (abstract).

4. Somerville PG, Greaves RL.Strong ground motions of the Kobe, Japan earthquake of January 17, 1995and development of a model of forward rupture directivity applicable in California. Proc. Western Regional Tech. Seminar on Earthquake Eng. for Dams. Assoc. of State Dam Safety Oficials, Sacramento, CA, April 11±12, 1996.

5. Abrahamson NA, Silva WJ.Empirical duration relations for shallow crustal earthquakes. Written communication, 1997

6. R. W. Graves, A. Pitarka, P. G. Somerville, Ground motion amplication in the Santa Monica area: effects of shallow basin edge structure. Submitted for publication.

7. P. G. Somerville, art. Emerging, ground. earthquake, In. motion, P. Dakoulas, M. Yegian, R. D. Holtz, editors, Proc Geotechnical Earthquake Engineering in Soil Dynamics III, Geotechnical Special Publication 751ASCE, 19981

8. Abrahamson NA.Spatial variation of multiple support inputs. Proc. First US Symp. Seism. Eval. Retrofit Steel Bridges, UC, Berkeley, October 18, 1993

9. F. Naeim, M. Lew, On the use of design spectrum compatible motions. Earthquake Spec 1995II(1):111±28.

10. Boulanger R.W. and Idriss, I.M.2006Liquefaction Susceptibility Criteria for Silts and Clays," Journal of Geotechnical and Geoenvironmental Engineering, 132 (11), 1413 EOF1426 EOFpp.

11. Boulanger R.W. and Idriss, I.M.2007Evaluation of Cyclic Softening in Silts and Clays", Journal of Geotechnical and Geoenvironmental Engineering, 133 (6), 641 EOF652 EOFpp.

12. Youd and Idriss, NCEER. Proceedings, Workshop on Evaluation of Liquefaction Resistance of Soils.Technical Report NCCER-97970022National Center for Earthquake

Engineering Research, University of Buffalo, Buffalo, New York, 1997

13. H. B. Seed, K. Tokimatsu, L. F. Harder, R. M. Chung, of. S. P. T. Influence, in. procedures, liquefaction. soil, evaluations. resistance, J Geotech Engng 19851425 EOF

14. Robertson PK, Fear CE.Liquefaction of sands and its evaluation. Proceedings, 1st Int. Conf. on Earthquake Geotechnical Engineering, Tokyo, Japan, 1995

15. Ambraseys NN. Engineering seismology.Earthquake Engng Struct Dynam 1988

16. Y. Wang, 2008aOn Contemporary Denotational Mathematics for Computational Intelligence. In Transactions of Computational Science (2629New York: Springer.

17. Y. Wang, 2009aOn Abstract Intelligence: Toward a Unified Theory of Natural, Artificial, Machinable, and Computational Intelligence. International Journal of Software Science and Computational Intelligence, 1 EOF17 EOF

18. Y. Wang, 2009bOn Cognitive Computing. International Journal of Software Science and Computational Intelligence, 1(3), 1-15.

19. A. M. Turing, 1950Computing Machinery and Intelligence. Mind, 59433460

20. Neumann. J. Von, 1946The Principles of Large-Scale Computing Machines. Reprinted in Annals of History of Computers, 3(3), 263-273.

21. Neumann. J. Von, 1958The Computer and the Brain. New Haven: Yale Univ. Press.

22. J. L. Gersting, 1982Mathematical Structures for Computer Science. San Francisco: W. H. Freeman & Co.

23. D. Mandrioli, C. Ghezzi, 1987Theoretical Foundations of Computer Science. New York: John Wiley & Sons.

24. H. R. Lewis, C. H. Papadimitriou, 1998Elements of the Theory of Computation, 2nd ed. Englewood Cliffs, NJ: Prentice Hall International.

25. J. Kephart, D. Chess, 2003The Vision of Autonomic Computing. IEEE Computer, 26(1), 41 EOF50 EOF

26. IBM2006Autonomic Computing White Paper. An Architectural Blueprint for Autonomic Computing, 4th ed., June, (137

27. Y. Wang, 2004Keynote: On Autonomic Computing and Cognitive Processes. Proc. 3rd IEEE International Conference on Cognitive Informatics (ICCI'04), Victoria, Canada, IEEE CS Press, (34

28. Y. Wang, 2007aJuly). Software Engineering Foundations: A Software Science Perspective. CRC Book Series in Software Engineering, Vol. II, Auerbach Publications, NY.

29. B. Bouchon-Meunier, R. Yager, L. Zadeh, 1995Fuzzy Logic and SoftComputing. World Scientific, Singapore.

30. J. C. Bezdek, is. a. What, intelligence?. computational, Zurada. J. M. In, I. I. R. J. Marks, C. J. Robinson, eds.) Computational Intelligence: Imitating Life, 112IEEE Press, Los Alamitos (1994

31. L. A. Zadeh, Sets. Fuzzy, Information and Control 8, 338-353 (1965

32. Zadeh L.A.The roles of fuzzy logic and soft computing in the conception, design and deployment of intelligent systems. BT Technol J. 14199443236

33. R. A. Aliev, R. R. Aliev, Computing. Soft, I. I. I. I. I. I. volumes, A. S. O. A. Baku, Press, 1997-1998in Russian).

34. R. Aliev, K. Bonfig, F. Aliew, Computing. Soft, Verlag. Berlin, Technic, 2000

35. Zadeh L.A. Foreword. In Proc.First European Congress on Intelligent Techniques and Soft Computing- EUFIT'95, page VII, 1995

36. L. A. Zadeh, Soft Computing and Fuzzy Logic. IEEE Software 11199464858

37. R. A. Aliev, Expert. Fuzzy, In. Systems, F. Aminzadeh, M. Jamshidi, S. O. F. T. C. O. M. P. U. T. I. N. G. (eds, Logic. Fuzzy, Networks. Neural, Artificial. Distributed, 9910 . Intelligence.pages, N. , NJ: PTR Prentice Hall, 1994

38. Zadeh L.A. Fuzzy logic, Neural Networks and Soft Computing .Comm of ACM 37199437784

39. Welstead S.T. (ed) Neural Networks and Fuzzy Logic Applications in C/C++, Professional Computing.NY: John Wiley, 1994

40. Yager R.R. and Zadeh L.A. (eds) Fuzzy sets, neural networks and Soft Computing.NY: VAN Nostrand Reinhold , 1994

41. Nauck D., Klawonn F., and Kruse R., Foundations of Neuro-Fuzzy Systems.NY: John Wiley and Sons, 1997.

42. Mohamad H.Hassoun, Fundamentals of artificial neural networks.Cambridge: MIT Press, 1995

43. S. Haykin, Networks. A. Neural, Foundation. Comprehensive, Marmillau, I. E. E. E. Computer, Society, 1994

44. Goldberg D.E., Genetic algorithms in search, optimization and machine learning.Reading, MA: Addison-Wesley, 1989

45. Arciszewski, and De Jong K.A. (2001Evolutionary computation in civil engineering: research frontiers. Eight International Conference on Civil and Structural Engineering Computing, (Topping B. H. V., ed.), Saxe-Coburg Publications, Eisenstadt, Vienna, Austria.

46. K. Miettinen, P.and. Neittaanmaki, J. Periaux, 1999Evolutionary Algorithms in Engineering and Computer Science : Recent Advances in Genetic Algorithms, Evolution Strategies, Evolutionary Programming, John Wiley & Sons Ltd., pps. 483. 0-47199-902-4

47. R. R. Youngs, S. J. Chiou, W. J. Silva, J. R. Humphrey, 1997Strong ground motion attenuation relationships for

subduction zone earthquakes. Seismol. Res. Lett., (68) 1, 58 EOF73 EOF

48. J. G. Anderson, 1997Nonparametric description of peak acceleration above a subduction thrust. Seismol. Res. Lett., (68) 1, 86 EOF93 EOF

49. C. B. Crouse, 1991Ground motion attenuation equations for earthquakes on the Cascadia subduction zone. Earth. Spectra, 7210236

50. S. K. Singh, M. Ordaz, M. Rodríguez, R. Quaas, V. Mena, M. Ottaviani, J. G. Anderson, D. Almora, 1989Analysis of near-source strong motion recordings along the Mexican subduction zone. Bull. Seism. Soc. Am., 7916971717

51. C. B. Crouse, Y. K. Vyas, B. A. Schell, 1988Ground motions from subduction-zone earthquakes. Bull. Seism. Soc. Am.,78125

52. S. K. Singh, E. Mena, R. Castro, C. Carmona, 1987Empirical prediction of ground motion in Mexico City from coastal earthquakes. Bull. Seism. Soc. Am., 7718621867

53. K. Sadigh, 1979Ground motion characteristics for earthquakes originating in subduction zones and in the western United States. Proc. Sixth Pan Amer. Conf., Lima, Peru.

54. B. F. Tichelaar, L. J. Ruff, 1993Depth of seismic coupling along subduction zones. J. Geophys. Res., 9820172037

55. G. M. Atkinson, D. M. Boore, 2003Empirical ground-motion Relations for Subduction-Zone Earthquakes and Their Applications to Cascadia and other regions. Bull. Seism. Soc. Am., 93417031729

56. S. C. Gómez, M. Ordaz, C. Tena, 2005Leyes de atenuación en desplazamiento y aceleración para el diseño sísmico de estructuras con aislamiento en la costa del Pacífico. Memorias del XV Congreso Nacional de Ingeniería Sísimica, México, Nov. A-II-02

57. D. Gasparini, E. H. Vanmarcke, 1976SIMQKE: A Program for Artificial Motion Generation, Department of Civil Engineering, Massachusetts Institute of Technology, Cambridge, MA.

58. W. J. Silva, K. Lee, 1987WES RASCAL code for synthesizing earthquake ground motions." State-of-the-Art for Assessing Earthquake Hazards in the United States, Report 24, U.S. Army Engineers Waterways Experiment Station, Misc. Paper S-731

59. B. A. Bolt, N. J. Gregor, 1993Synthesized Strong Ground Motions for the Seismic Condition Assessment of the Eastern Portion of the San Francisco Bay Bridge", Report UCB /EERC-93/12, University of California, Earthquake Engineering Research Center, Berkeley, CA.

60. J. E. y. C. A. Carballo, Cornell, 2000Probabilistic seismic demand analysis: spectrum matching and design", Department of Civil and Environmental Engineering, Stanford University, Report RMS-41

61. C. Kircher, 1993Personal communication with Farzad Naeim and Marshall Lew.

62. F. Naeim, J. Kelly, 1999Design of Seismic Isolated Structures from Theory to Practice. New York, John Wiley & Sons. 289p.

63. N. E. Huang, S. Zheng, S. R. Long, M. C. Wu, H. H. Shih, Q. Zheng, N. Yen, C. , C. C. Tung, M. H. Liu, 1998The empirical mode decomposition and Hilbert spectrum for nonlinear and nonstationary time series analysis", Proc. R. Soc. London, Ser. A 454903995

64. A. Roulle, F. J. Chavez-Garcia, 2006The strong ground motion in Mexico City: Analysis of data recorded by a 3D array, Soil. Dyn. Eq. Eng. 2671

65. H. Kawase, K. Aki, 1989A study of the response of a soft bas in for incident S, P and Rayleigh waves with special reference to the long duration observed in Mexico City, Bull Seism. Soc. Am. 7913611382

66. N. Yoshida, S. Iai, 1998Nonlinear site response and its evaluation and prediction," IN Irikura, K., Kudo, K., Okada, K. & Sasatani, T. (Eds.) The Second International Symposium on the Effects of Surface Geology on Seismic Motion, Yokohama, Japan, A.A.Balkema, 7190

67. I. y. Herrera, E. . Rosenblueth, Response Spectra on Stratified Soil". Proc. 3rd. World Conference on Earthquake Engineering. Nueva Zelandia, 44561965

68. M. P. Romo, A. Jaime, 1986Dynamic characteristics of some clays of the Mexico Valley and seismic response of the ground". Technical Report, Apr., Instituto de Ingenieria, Mexico City, Mexico (in Spanish).

69. A. Jaime, M. P. Romo, D. Reséndiz, 1988Comportamiento de pilotes de fricción en arcilla del valle de México." Series of the Instituto de Ingeniería, Mexico City, Mexico

70. M. P. Romo, Seed, 1986Analytical modelling of dynamic soil response in the Mexico Earthquake of September 19, 1985". Proc. ASCE Int. Conf. on the Mexico Earthquakes-1985148162

71. S. R. García, M. P. Romo, J. Mayoral, 2007Estimation of Peak Ground Accelerations for Mexican Subduction Zone Earthquakes using Neural Networks", Geofísica Internacional, 46-15163enero-marzo

72. G. Castro, S. J. Poulos, J. W. France, J. L. Enos, 1982Liquefaction induced by cyclic loading. Winchester, Mass: Geotechnical Engineers Inc.

73. H. B. Seed, I. M. Idriss, I. Arango, . Evaluation, Liquefaction. of, Using. Potential, Performance. Field, Data, of. Journal, Geotechnical. the, Division. A. S. C. E. Engineering, Vol, GT31983Seed et al., 1983

74. H. B. Seed, I. Idriss, M.1971Simplified procedure for evaluation soil liquefaction potential. Journal of the Soil Mechanics and Foundations ASCE, 97 (9), 1249-1273.

75. T. L. Youd, I. M. . Idriss, R. D. Andrus, I. Arango, G. Castro, J. T. Christian, R. Dobry, F. Liam, L. F. Harder, M. E. Hynes, K. Ishihara, J. P. Koester, S. S. C. Liao, I. I. I. W. F. Marcuson, G. R. Martin, J. K. Mitchell, Y. Moriwaki, M. S. Power, P. K. Robertson, R. B. Seed, K. H. Stokoe, 200, Liquefaction resistance of soils. Summary report from the 1996NCEER and 1998 NCEER/NSF workshops on evaluation of liquefaction resistance of soils. J. Geotech. Geoenviron. Eng., 127(10), 817-833.

76. R. Boulanger, I. M. Idriss, 2004State normalization of penetration resistance and the effect of overburden stress on liquefaction resistance. Proc. 11th International Conf. on Soil Dynamics and Earthquake Engineering and 3rd International Conference on Earthquake Geotechnical Engineering, Univ. of California, Berkeley, CA.

77. S. M. Fatemi-Agdha, M. Teshnehlab, A. Suzuki, T. Akiyoshi, Y. Kitazono, 1998Liquefaction potential assesment using multilayer artificial neural network. J. Sci. I.R. Iran, 9(3).

78. C. H. Juang, C. J. Chen, Y. M. Tien, 1999Appraising cone penetration test based liquefaction resistance evaluation methods: Artificial neural networks approach. Canadian Geotechnical Journal, 363443454

79. C. H. Juang, H. M. Yuan, D. H. Lee, Lin, 2003P. S., "Simplified cone penetration test-based method for evaluating liquefaction resistance of soils," Journal of Geotechnical and Geoenvironmental Engineering, 12916680

80. M. H. Baziar, N. Nilipour, 2003Evaluation of liquefaction potential using neural-networks and CPT results. Soil Dynamics and Earthquake Engineering, 237631636

81. S. Chern, C. Lee, 2009CPT-based simplified liquefaction assessment by using fuzzy-neural network. Journal of Marine Science and Technology, 174326331

82. García S.R. and Romo M.P.2007GENES: Genetic Generator of Signals, a Synthetic Accelerograms Application. , Proc. of the SEE5, SM-80.

83. S. F. Bartlett, T. L. Youd, 1992Empirical Analysis of Horizontal Ground Displacement Generated by Liquefaction Induced Lateral Spreads", Tech. Rept. NCEER 920021National Center for Earthquake Engineering Research, SUNY- Buffalo, Buffalo, NY.

84. Youd T.L., Hansen C.M. and Bartlett S.F.2002Revised Multilinear Regression Equations for Prediction of Lateral Spread Displacement' Journal of Geotechnical and Geoenvironmental Engineering, 1281210071017

2

CHAPTER

Modifying the Intensity Distribution by Assessing the Reliability

Lan-chi Kang • Xing Jin • Yong-xiang Wei

Earthquake Administration of Fujian Province, Fuzhou 350003, China

ABSTRACT

This article presents an application of a procedure to modify the intensity distribution by assessing the reliability. There are two potential possibilities that may influence the intensity distribution: (1) For the interpolation error, we generate a measured grid across the calculation region. When the point to station spacing is <5 km, we consider the results precise; however, some points have less precision because these are farther from the corresponding stations. When the spacing is between 5 and 50 km, we consider the results imprecise and define a reliability factor that correlates with the distance. (2) Some records may have errors that result from local site conditions, equipment problems, or some disturbance such as lightning stroke, which will lead to some grid points having an incorrect intensity. We regress the attenuation relation for sites with abnormal intensities and consider the results to be accurate when the standard deviation (STD) is $<\sigma$ and inaccurate when the STD is $> 2\sigma$. We then define a reliability factor to correlate with STD between σ and 2σ, such that the intensity distribution is in accord with both wave propagation theory and the investigation intensity.

Keywords ReliabilityIntensity distributionInterpolation errorStation error Wenchuan earthquakeChi–Chi earthquake

1 INTRODUCTION

For rapid emergency response, such as resource dispatch management, information regarding the spatial distribution of seismic intensity is particularly useful, and a near real-time or quick damage assessment in terms of seismic intensity is greatly in demand. One practical way to fulfill this demand is to correlate seismic intensity with peak ground motions because they can be acquired in near real time via a rapid reporting system (Lind et al. 1996; Yamazaki et al.1997; David et al. 1999; Kanamori 1993; Ma et al. 2003; Jin et al. 2008; Wei et al. 2010).

On May 12, 2008, the large magnitude Wenchuan earthquake struck the densely populated Sichuan Province of China (31.0°N, 103.4°E). More than 80,000 people were killed, ~370,000 people were injured, and economic losses have been estimated to be in excess of 800 billion RMB. The large moment magnitude (M_W 7.9) (Hao et al. 2009; Shen Sun et al. 2009; Xu et al.2009; Zhang et al. 2009) of the event attests to it being a great intercontinental earthquake. The Wenchuan earthquake occurred on the Longmen Shan fault zone, which delineates the eastern margin of the Tibet Plateau. Located on the western circum-Pacific seismic belt, Taiwan is situated in the collision zone between the Philippine Sea plate (PSP) and the Eurasian plate (EP). As a result, seismicity in Taiwan is considerably high, as demonstrated by the catastrophic 1999 Jiji (Chi–Chi) earthquake. For rapid reporting of felt earthquakes within Taiwan, a real-time strong motion network, the Taiwan Rapid Earthquake Information Release System (TREIRS), has been operated by the Central Weather Bureau (CWB) since 1997 (Shin and Teng2001; Wang et al. 2005). In addition to routine magnitude and hypocenter determinations following a felt earthquake, peak ground motions recorded by the TREIRS system are used to generate seismic intensity.

In this paper, we use the records from the Wenchuan and Jiji (Chi–Chi) earthquakes to simulate the progress of rapid reporting of seismic intensity. After mesh generation in the calculation region, we apply the interpolation method to calculate the peak ground acceleration (PGA) for each grid. In accordance with the intensity scale, we transform the PGA to intensity. However, owing to some problems, following the analysis of the intensity distribution, we present a reliability index to modify the intensity distribution.

2 PGA AND INTENSITY

In general, PGA is calculated in the reporting system of seismic intensity; however, estimating the intensity from PGA is a complex academic problem. Therefore, in this paper, we analyze the relation between intensity and PGA, which are given on the Chinese seismic intensity scale (GB/T 17742-2008). The relation can be written as

$$I = a \ln A_m + b, \tag{1}$$

where Am is the PGA, I is the intensity, and *a* and *b* are the regressed coefficients.

Figure 1 shows the relation between A_m and I. In the areas below I = IV, PGAs corresponding to intensity are not given, and the regression is shown as a dash–dot line. Based on this result, we can directly calculate the intensity from PGA.

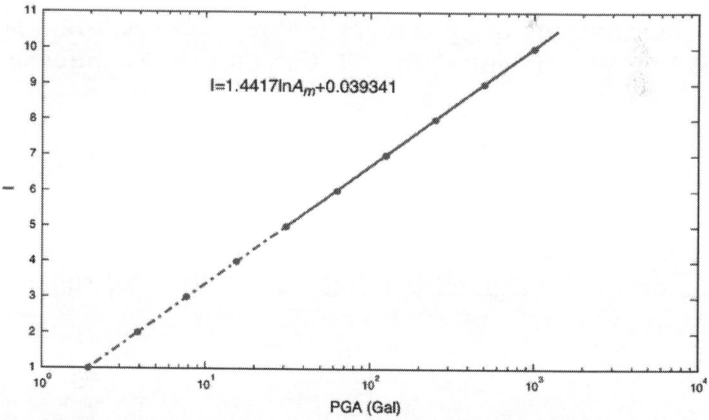

Fig. 1 The relation between A_m and I. The PGA corresponding to intensity could be calculated by the relation if I < IV

3 THE INTENSITY DISTRIBUTION

In this study, we divide the calculated regions into 16,416 grids with a grid spacing of 0.02° and 21,091 grids with a grid spacing of 0.05° for the Jiji (Chi–Chi) and Wenchuan earthquakes, respectively. From the catalog, 379 records distributed across the whole of Taiwan have been selected for the Chi–Chi earthquake (Fig. 2b), and 87 records have been selected from the Wenchuan earthquake, including all stations in Sichuan Province and additional stations in Gansu and Shanxi provinces (Fig. 2a).

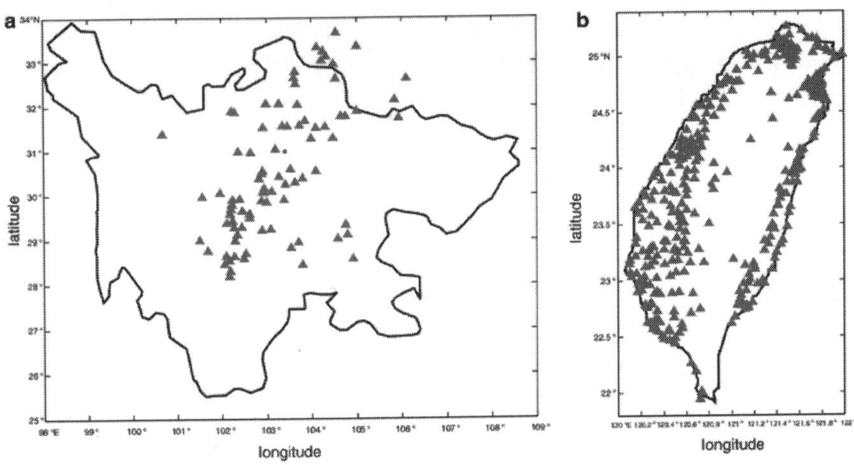

Fig. 2 Distributions of epicenters (*circle*) and recording stations (*triangle*) for Wenchuan (**a**) and Jiji (Chi–Chi) (**b**) earthquake

Peak ground acceleration and peak ground velocity (PGV) grid points are calculated by an interpolation methodology as shown in Fig. 3 (Wei et al. 2010; Zhang 2008). In this interpolation methodology, the weighted is refined as the inverse square of the distance, as follows:

$$
\begin{cases}
W_a = \dfrac{\left(\frac{1}{\Delta_{ad}}\right)^2}{\left(\frac{1}{\Delta_{ad}}\right)^2 + \left(\frac{1}{\Delta_{bd}}\right)^2 + \left(\frac{1}{\Delta_{cd}}\right)^2} \\
W_b = \dfrac{\left(\frac{1}{\Delta_{bd}}\right)^2}{\left(\frac{1}{\Delta_{ad}}\right)^2 + \left(\frac{1}{\Delta_{bd}}\right)^2 + \left(\frac{1}{\Delta_{cd}}\right)^2} \\
W_c = \dfrac{\left(\frac{1}{\Delta_{cd}}\right)^2}{\left(\frac{1}{\Delta_{ad}}\right)^2 + \left(\frac{1}{\Delta_{bd}}\right)^2 + \left(\frac{1}{\Delta_{cd}}\right)^2}
\end{cases}
\quad (2)
$$

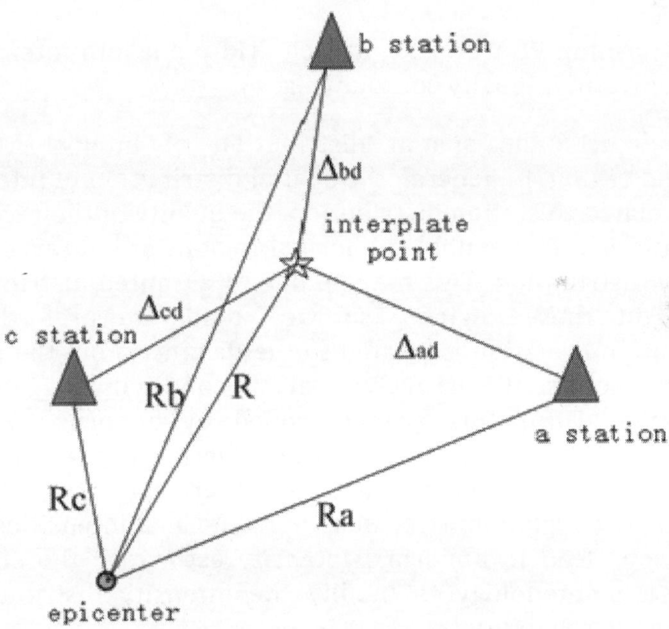

Fig. 3 Sketch of relation between station, interpolate points, and epicenter

The methodology can be written as follows:

$$A(R) = W_a A_a f(R, R_a) + W_b A_b f(R, R_b) + W_c A_c f(R, R_c) \tag{3}$$

where Wa,Wb,Wc is the weighted, Aa,Ab,Ac is the PGA of *a*, *b*, and *c* stations, and f(R,Ra),f(R,Rb),f(R,Rc) is the coefficient of attenuation relations.

Table 1 Attenuation relation

Earthquake		C_1	C_2	σ
Wenchuan	PGA	8.1946	−0.59755	0.51101
Jiji (Chi–Chi)	PGA	7.9001	−0.82078	0.60696

By transforming PGA or PGV at each grid point into intensity, we can generate an intensity contour (Fig. 4).

Figure 4 depicts the spatial distribution of intensity for the calculated region. In general, the spatial distribution of intensity is closely related to station distribution, the quality of the records, or site conditions. Two notable abnormal regions are observed in the intensity distribution. This may be due to a limited distribution of stations in these regions, so we could interpret that the interpolation precision is related to the distance from the stations to the grid points. Alternatively, local site conditions may influence PGA. Some grid points are controlled by abnormal PGA, for example, Fig. 5 showing three records, where the PGAs are higher contrast to some records with a closer epicentral distance; therefore, the region must be showing abnormal intensities. These two reasons lead to abnormal intensity estimates; therefore, we present a methodology to modify the intensity distribution by assessing the reliability.

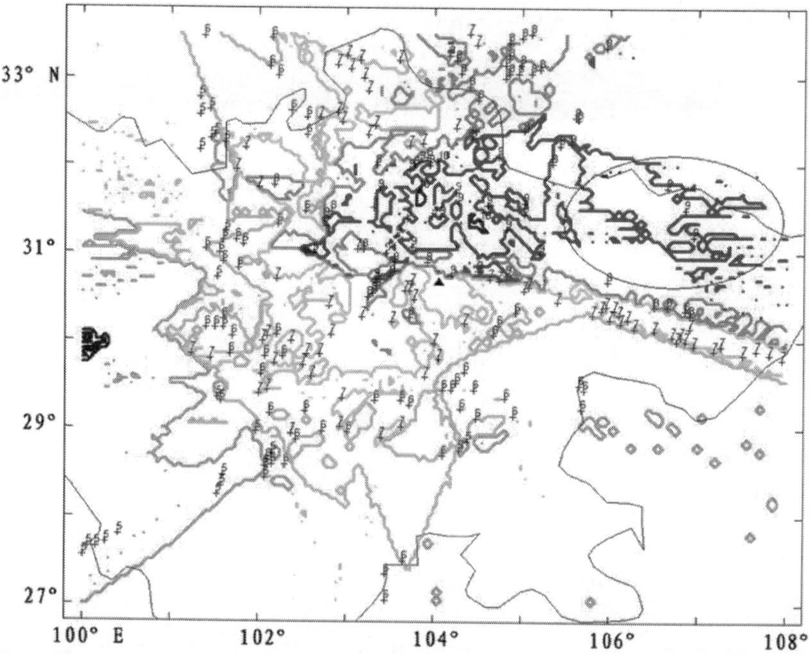

Fig. 4 Intensity distribution map. The notable abnormal regions are observed in the intensity distribution. This may be due to station distribution, the quality of the records, or site conditions

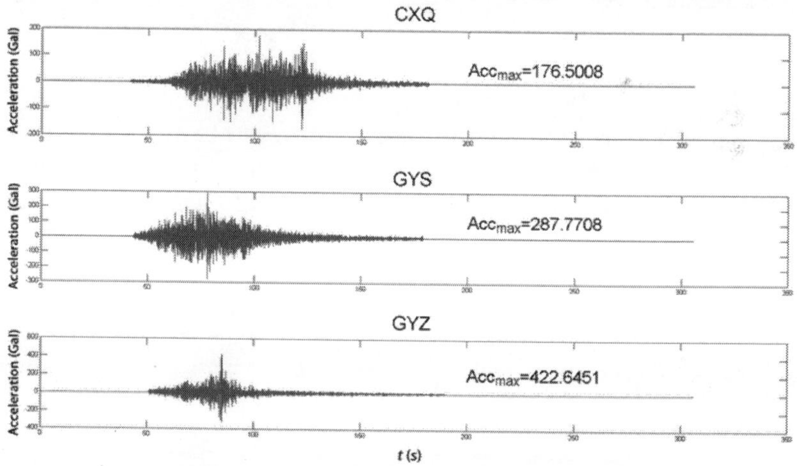

Fig. 5 Stations with abnormal PGA. The PGAs are higher contrast to some records with a closer epicentral distance; therefore, the region must be showing abnormal intensities

4 RELIABILITY ASSESSMENTS

4.1 Reliability of the attenuation relation of PGA

The most commonly used attenuation model in China that is recommended for*EvaluationofSeismicSafetyforEngineeringSites* (GB17741-2005) is

$$\log Y = C_1 + C_2M + C_3M^2 + C_4\log(R + R_0(M)) + C_7R + \sigma, \tag{4}$$

where *Y* is the peak ground motion or response spectrum at a variable period; Ci (i = 1, 2, ...7) is the regression coefficient; *M* is the local magnitude; *R* is the fault distance, which can be written as *D*; R0(M)=C5exp(C6M) is a saturation factor that has a correlation with the earthquake intensity. In general, R0(M) is the hypocentral depth and σ is the random variable. The C2M+C3M2 is a constant if the attenuation is response to particular earthquake. In general, the coefficient C7R is very small, which shows that the coefficient C7 has less relation to Y; therefore, we remove C7R, so that Eq. (3) can be written as

$$\log Y = C_1 + C_2\log(D + 10) + \sigma \tag{5}$$

Figure 6 and Table 1 shows the regression results and the standard deviation (STD) of PGA. We generally accept the PGA when the STD < σ and assume that stations with STD > 2σ are not valid; thus, the PGA reliability can be refined as

$$P(A_m) = \begin{cases} 0 & \eta_A \geq 2\sigma_A \\ \frac{2\sigma_A - \eta_A}{\sigma_A} & \sigma_A \leq \eta_A < 2\sigma_A \\ 1 & \eta_A < \sigma_A \end{cases} \tag{6}$$

where η is residuals between observation and the attenuation relation, σ is the STD, $P(A_m)$ is the PGA reliability, and the modifying methodology can be written as:

$$A'_m(x, y|A_m) = P(A_m)A_m + [1 - P(A_m)]\bar{A} \quad (7)$$

where A_m is the observation PGA, $\bar{A}$ is the calculated PGA by the attenuation relation.

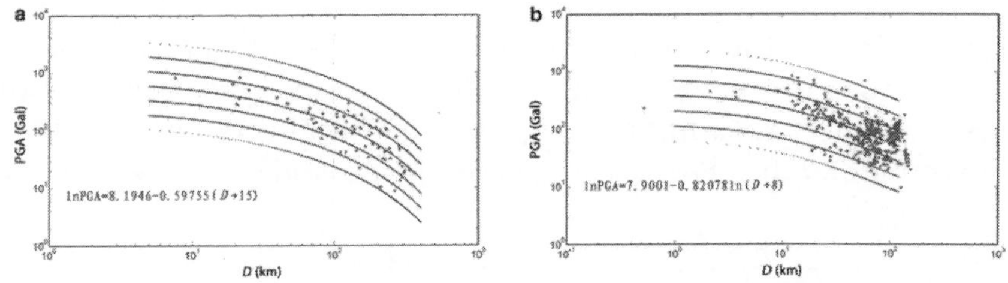

Fig. 6 PGA regression results. **a** Wenchuan earthquake; **b** Jiji (Chi-Chi) earthquake. The area between black lines is defined as 3σ; the area between red lines is defined as 2σ; the area between blue lines is defined as σ

4.2 **Reliability of the interpolation distance**

We use an interpolation method of inverse weighted square distance, presented by Jin and Zhang, where the interpolation precision is related to the interpolation distance. This method is very accurate when the distance from point to the station is <5 km (Rmin). However, if the distance from point to the station is >50 km (Rmax), the method is inaccurate and these grid points will be calculated by the attenuation relation, and we define a reliability index with which the distance correlates, when the distance is between 5 and 50 km. The reliability P(R) of interpolation distance can be refined as:

$$P(R) = \begin{cases} 1 & R \leq R_{\min} \\ 1 - \frac{R - R_{\min}}{R_{\max} - R_{\min}} & R_{\min} < R \leq R_{\max} \\ 0 & R > R_{\max} \end{cases} \quad (8)$$

The modifying intensity distribution can be written as:

$$A'_m = P(R)A_m + [1 - P(R)]\bar{A}, \quad (9)$$

where A′m is the modified PGA, Am is the unmodified PGA, and $\bar{A}$ is calculated by the attenuation.

4.3 **Modifying method of the synthetic reliability**

The synthetical reliability can be refined as:

$$P_s = P(R) * P(A_m). \tag{10}$$

Figure 7 depicts the spatial distributions of the synthetic reliability; we find that the error regions of previous intensity distribution have a very low reliability, so we confirm that this method could be used to modify the errors. The modifying method can be written as:

$$A'_m = P_s A_m + (1 - P_s)\bar{A} \tag{11}$$

where A′m is the modified PGA, Am is the unmodified PGA, and A¯ is calculated by the attenuation Ps is the synthetic reliability.

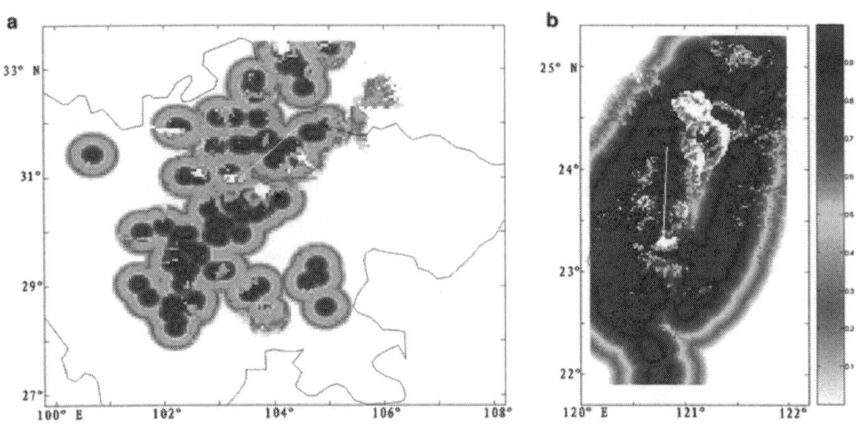

Fig. 7 Distribution of synthetic reliability. **a** Wenchuan earthquake; **b** Jiji (Chi–Chi) earthquake. The error regions of previous intensity distribution have a very low reliability, so these errors could be modified

5 **RESULTS**

Figure 8 shows the modified intensity distribution, which can be differentiated from Fig. 3 by the absence of regions with abnormal intensities, and the grid points in these regions are calculated by the attenuation relation. This methodology mainly modifies the points with greater separation between station and points, with the aid of which the results accord not only with the theory of the wave propagation, but also with the investigation intensity.

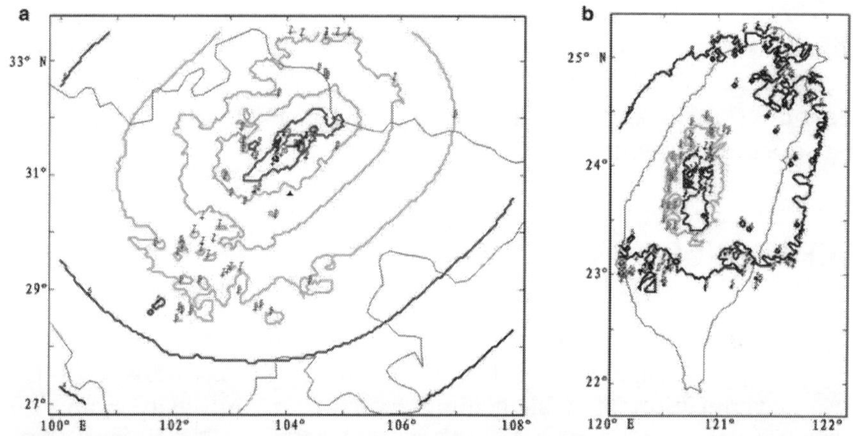

Fig. 8 Modified intensity distribution for the study areas. **a** Wenchuan earthquake; **b** Jiji (Chi–Chi) earthquake. There are some differences between Figs. 3 and 8, and some grid points with greater separation are modified

6 DISCUSSION AND CONCLUSIONS

We analyze the intensity distribution of the Wenchuan and Jiji (Chi–Chi) earthquakes. We found that some interpolation errors and abnormal stations influence the intensity distribution. First, we find that some stations with an abnormal PGA have affects on the intensity distribution and introduce the reliability of PGA, refining the PGA reliability as STD, which are calculated by the attenuation relation. We then introduce the interpolation reliability; however, some researchers suggest that the interpolation methodology has only high precision if the distance is less than 5 km, but gives large errors when the distances are greater than 50 km. So we adopt the PGA attenuation relation to modify some error grid points in the intensity distribution. The modified results are in accord with wave propagation theory and also meet the investigation intensity. All of these tasks can be completed in 2 min; thus, this methodology can be applied to the rapidly reporting system of seismic intensity.

OPEN ACCESS

REFERENCES

1. David JW, Quitoriano V, Heaton TH, Kanamori H, Scrivner CW, Worden CB (1999) TriNet "ShakeMaps": Rapid generation of peak ground motion and intensity maps for earthquake in Southern California. Earthq Spectra 15:537–556

2. General administration of quality supervision, inspection and quarantine of the People's Republic of China (2008) The Chinese seismic intensity scale (GB/T 17742-2008). China Zhijian Publishing House, Beijing

3. Hao KX, Si HJ, Fujiwara H, Ozawa T (2009) Coseismic surface ruptures and crustal deformations of the 2008 Wenchuan earthquake M_W 7.9, China. Geophys Res Lett 36:L11303

4. Jin X, Zhang HC, Wei YX (2008) Shaking Intensity Criterion Based on Rapid Release Information by Seismic Monitoring Network and Its Application. Recent Dev World Seismol 10:20–27 (in Chinese with English abstract)

5. Kanamori H (1993) Location earthquake with amplitude application to real-time seismology. Bull Seismol Soc Am 83:264–268

6. Lind SG, Douglas SN, Douglas SD, Michael EP (1996) Real-time seismology at UC Berkeley: the rapid earthquake integration project. Bull Seismol Soc Am 86:936–945

7. Ma Q, Jin X, Li SY (2003) Real-time algorithm of seismic dynamic response for SDOF system. Earthq Eng Eng Vib 23:61–68 (in Chinese with English abstract)

8. Shen Sun Z, Zhang J, Wan P, Wang Y, Bürgmann M, Zeng R, Gan YW, Liao H, Wang Q (2009) Slip maxima at fault junctions and rupturing of barriers during the 2008 Wenchuan earthquake. Nat Geosci 2(10):718–724

9. Shin TC, Teng TL (2001) An overview of the 1999 Chi–Chi Taiwan earthquake. Bull Seismol Soc Am 91:895–913

10. Wang WM, Zhao LF, Li J, Yao ZX (2005) Rupture process of the Chichi (Taiwan) earthquake in 1999. Chin J Geophys 48:132–147 (in Chinese with English abstract)

11. Wei YX, Li J, Jin X, Zhang HC (2010) Preliminary study of near real-time interpolation calculation of ShakeMap based on seismic monitoring networks information. J Disaster Prev Reduct 26:2–11 (in Chinese with English abstract)

12. Xu X, Wen X, Yu G, Chen G, Klinger Hubbard YJ, Shaw J (2009) Coseismic reverse- and oblique-slip surface faulting generated by the 2008 M_W 7.9 Wenchuan earthquake, China. Geology 37:515–518

13. Yamazaki F, Noda S, Meguro K (1997) Developments of monitoring and early damage assessment system in Japan. Bull ERS 30:15–31

14. Zhang HC (2008) Research on ShakeMaps and Shaking Intensity Based on Records for Monitoring Network. Master Dissertation, Harbin: Institute of Engineering Mechanics State Seismological Bureau, China (in Chinese)

15. Zhang Y, Feng WP, Xu LS, Zhou CH, Chen YT (2009) Spatio-temporal rupture process of the 2008 great Wenchuan earthquake. Sci China Earth Sci 52:145–154

3

CHAPTER

Effect of Constitutive Material Models on Seismic Response of Two-Story Reinforced Concrete Frame

Md. Iftekharul Alam and Dookie Kim

Department of Civil and Environmental Engineering, Kunsan National University, Kunsan 573-701, Korea.

ABSTRACT

This paper focuses on the finite element (FE) response sensitivity and reliability analyses considering smooth constitutive material models. A reinforced concrete frame is modeled for FE sensitivity analysis followed by direct differentiation method under both static and dynamic load cases. Later, the reliability analysis is performed to predict the seismic behavior of the frame. Displacement sensitivity discontinuities are observed along the pseudo-time axis using non-smooth concrete and reinforcing steel model under quasi-static loading. However, the smooth materials show continuity in response sensitivity at elastic to plastic transition points. The normalized sensitivity results are also used to measure the relative importance of the material parameters on the structural responses. In FE reliability analysis, the influence of smoothness behavior of reinforcing steel is carefully noticed. More efficient and reasonable reliability estimation can be achieved by using smooth material model compare with bilinear material constitutive model.

Keywords: finite element analysissensitivity analysisstructural reliabilityconstitutive modelsstructural response

1 INTRODUCTION

The philosophy of performance-based earthquake engineering has gained recognition widely in the structural analysis and design field and has been included in many seismic design guidelines (ATC-55 2005; BSSC 2003) with noticeable advances in the field of structural reliability since last two decades. Several numerical and analytical methodologies have been developed considering the non-linear behavior, material parameters, and geometric uncertainties (Schueller et al. 2004). Der Kiureghian and Ke (1988) successfully implemented reliability methodology by using finite element (FE) method and first-order reliability method (FORM) (Ditlevsen and Madsen 1996). These methods were examined to investigate the influence of the correlation length of random property or load fields on the reliability of the subjects. The results have shown that the correlation length of property fields has influence on the displacement responses, but it may not be significant in stress limit states. The determination of design point(s) is considered as an essential step in gradient based reliability methodology. Structural response sensitivity is an important ingredient and a by-product of the design point search (Hohenbichler and Rackwitz 1986). For this purpose, the use of direct differentiation method (DDM) seems to be efficient and accurate approach to perform sensitivity analysis. A method has been presented to get key sensitive attributes in the material constitutive and discrete loading parameters of force-based FE frame systems (Conte et al. 2004). The material non-linearity was considered to get the static and dynamic responses using DDM. It was concluded that the developed procedure prone to sensitivity offers a powerful general tool for computing the responses. The analytical procedure and guidelines were developed in OpenSees (Mazzoni et al. 2005; Gu et al. 2010) framework to perform FE based reliability analysis (Haukaas and Der Kiureghian 2004). In this context, smoothed materials were modeled and the existing search algorithms were modified consequently. In case of non-linear beam column elements, response sensitivities have been evaluated with great precision using DDM (Scott et al. 2004; Scott and Filippou 2007). Similarly, the application of sensitivity and reliability analyses in the soil-foundation-structure-interaction systems has been reported in several research articles (Gu2008; Gu et al. 2009a, b). Scott (2012) evaluated two existing formulations of force-based element response sensitivity and found the consistency of first formulation, while the second one proved to be inconsistent with a high condition number.

For gradient based methods, sensitivity analysis is performed to find the design point in reliability analysis. The continuity of the obtained response is essential for most of the algorithms. The discontinuity increases computational effort and due to using improper material model the process can be interrupted. Haukaas and Der Kiureghian (2004) performed static analysis on single degree of freedom systems to explain the discontinuity effects in the displacement sensitivity results using smooth and non-smooth material models. The results have shown the occurrence of discontinuity at yielding and unloading points in the bi-linear material. It has been suggested that by using the smooth material models such effect can be avoided. Another study was conducted focusing on the effect of gradient discontinuities caused by non-smoothness of the material models in the reliability context (Barbato and Conte 2006). A structural system was modeled using smooth and non-smooth reinforcing steels to compare response sensitivity and reliability analysis results. The consistency of the results was fairly verified with these obtained by previous researchers. Furthermore, study on the dynamic analysis concluded that the discontinuities can be effectively eliminated by providing smoothing effect in reinforcing steel and refining the time discretization of the equations of motion.

This paper presents the modeling and analysis of a reinforced concrete (RC) structure using different concrete and steel constitutive material models in the context of sensitivity based reliability analysis. DDM based response sensitivity analysis has been conducted for smooth and non-smooth material constitutive laws. Quasi-static cyclic analysis results show that discontinuities arise in the response sensitivities while using non-smooth material models. Moreover, this difficulty can be eliminated by using smooth constitutive material models. Later, the smoothness effect of reinforcing steel on failure probabilities of frame structure has been observed using structural reliability analysis.

2 MATERIAL MODELS

Different concrete and steel material constitutive models are used to describe the structural behavior considering smoothness effect of the materials. The Kent–Scott–Park (Scott et al.1982) concrete model is adopted as zero tension and convenient for application in engineering for its simple form. While Smoothed Popovics–Saenz (Balan et al. 1997, 2001; Zona et al. 2005) concrete also modeled with zero tension stiffening having smoothed skeleton curve. In FE structural modeling, fiber concrete sections using confined and unconfined concrete layer are considered. Figures 1 and 2 show the typical comparison between two concrete models. In case of steel material, 1D J_2 plasticity model (Conte et al. 2003) (also known as bi-linear inelastic model) and smooth inelastic Menegotto–Pinto (M–P) model (Menegotto and Pinto 1973) are used in material modeling of the structure as shown in Fig. 3. Bi-linear model with pure kinematic hardening, H_{kin} (zero isotropic hardening, H_{iso}) is considered. As a non-smooth plasticity model, J_2 plasticity model with Von Mises yield surface is well-known for metallic materials. This model has shown discontinuities in the responses when the material enters the plastic regime (Conte 2001). The version of M–P model extended by Filippou et al. (1983) has been considered to account for the isotropic strain hardening. Figure 3 shows that yield stress, f_y and initial elastic tangent (modulus of elasticity), E for both of the steel models are same. Post yield tangent modulus for J_2 plasticity model and M–P model can be expressed by the Eqs. (1) and (2) as follows:

$$E_P = \frac{E(H_{kin} + H_{iso})}{E + H_{kin} + H_{iso}} \tag{1}$$

$$E_P = b \times E \tag{2}$$

where b is known as strain-hardening ratio and R_0 controls the transition from elastic to plastic regions (Fig. 3). The M–P 1D plasticity model has shown good agreement with the experimental results, particularly for cyclic tests on reinforcing steel bars. It is observed that the constitutive law of M–P model is smooth and continuously differentiable with respect to constitutive material parameters, thus response sensitivities continuous everywhere. Local buckling effect of steel bars in RC members can be taken into account by further modifications in M–P model (Monti and Nuti 1992).

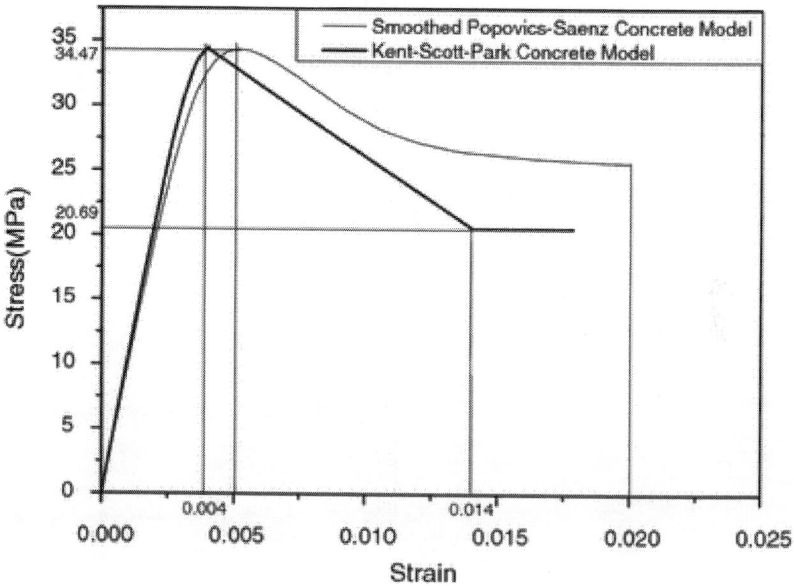

Fig. 1 Comparison between two confined concrete models.

3 DDM BASED SENSITIVITY ANALYSIS

FE response sensitivity analysis is an important analytical tool in structural optimization, FE model updating, reliability analysis, and structural identification (Ditlevsen and Madsen 1996; Kleiber et al. 1997). The response sensitivity is a measure of the change in the response quantity due to a unit change in a system parameter. Cross-sectional geometry, materials properties, applied loads or nodal co-ordinates are considered as the key parameters. Generally, the structural response can be characterized in terms of

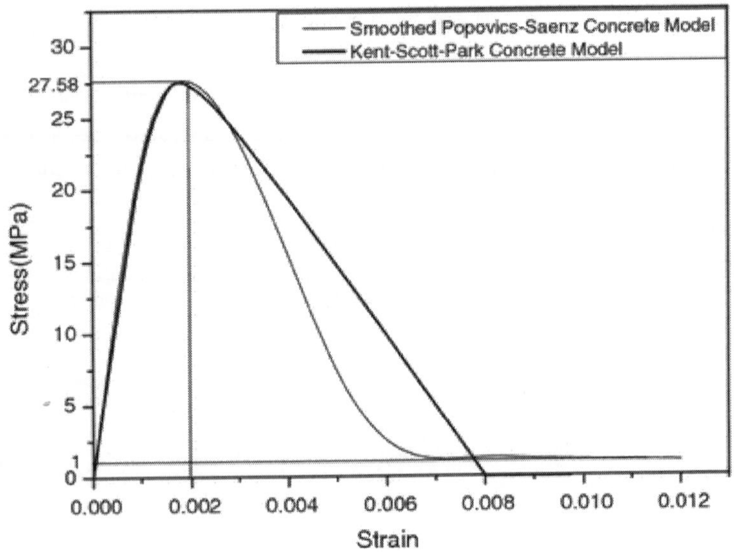

Fig. 2 Comparison between two unconfined concrete models.

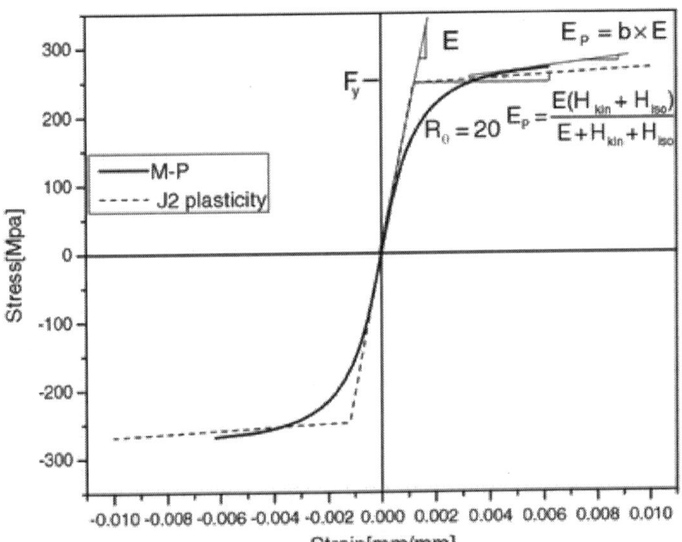

Fig. 3 Typical stress–strain curves of different constitutive steel models.

deformations, forces, or integrated quantities, i.e., dissipated energy and accumulated damage (Haukaas and Der Kiureghian2004). Such response can be computed by structural analysis tools such as OpenSees. The DDM method is well known for computing response quantities with high level of accuracy (Kleiber et al. 1997; Arora and Haug 1979; Zhang and Der Kiureghian 1993). In nonlinear DDM based analysis structural response is calculated in each time step after achieving the convergence of the response computation. Differentiation of the algorithm is required to obtain the response using specific sensitivity parameter.

The equations of the response sensitivities were derived by Zhang and Der Kiureghian (1993). After spatial discretization using the FE method, the equation of motion of the considered system can be expressed as (Gu et al. 2009b):

$$\boldsymbol{M}(\theta)\ddot{\boldsymbol{u}}(t,\theta) + \boldsymbol{C}(\theta)\dot{\boldsymbol{u}}(t,\theta) + \boldsymbol{R}[\boldsymbol{u}(t,\theta),\theta] = \boldsymbol{F}(t,\theta) \qquad (3)$$

where t is the time, θ is the scalar sensitivity parameter as a material constitutive parameter, $\boldsymbol{u}$is the vector of nodal displacements, $\boldsymbol{M}$ is the mass matrix, $\boldsymbol{C}$ is the damping matrix, $\boldsymbol{R}$ is the history dependent internal (inelastic) resisting force vector, and $\boldsymbol{F}$ is the applied dynamic load vector.

Newmark-β method is used for the numerical integration of Eq. (3). The dynamic residual $\Psi(\boldsymbol{u}_{n+1})$ can be written at discrete time $t = t_{n+1} = (n+1)\Delta t$(where Δt is the constant integration time step) as follows:

$$\Psi(\boldsymbol{u}_{n+1}) = \tilde{\boldsymbol{F}}_{n+1} - \left[\frac{1}{\beta(\Delta t)^2}\boldsymbol{M}\boldsymbol{u}_{n+1} + \frac{\alpha}{\beta(\Delta t)}\boldsymbol{C}\boldsymbol{u}_{n+1} + \boldsymbol{R}(\boldsymbol{u}_{n+1})\right] = 0 \qquad (4)$$

$\tilde{\boldsymbol{F}}_{n+1}$ can be expressed as below:

$$\tilde{\boldsymbol{F}}_{n+1} = \boldsymbol{F}_{n+1} + \boldsymbol{M}\left[\frac{1}{\beta(\Delta t)^2}\boldsymbol{u}_n + \frac{1}{\beta(\Delta t)}\dot{\boldsymbol{u}}_n - \left(1 - \frac{1}{2\beta}\right)\ddot{\boldsymbol{u}}_n\right] + \boldsymbol{C}\left[\frac{\alpha}{\beta(\Delta t)}\boldsymbol{u}_n - \left(1 - \frac{\alpha}{\beta}\right)\dot{\boldsymbol{u}}_n - (\Delta t)\left(1 - \frac{\alpha}{2\beta}\right)\ddot{\boldsymbol{u}}_n\right]$$

Differentiating Eq. (4) with respect to θ,

$$\left[\frac{1}{\beta(\Delta t)^2}\boldsymbol{M} + \frac{\alpha}{\beta(\Delta t)}\boldsymbol{C} + \left(\boldsymbol{K}_T^{stat}\right)_{n+1}\right]\frac{\partial \boldsymbol{u}_{n+1}}{\partial \theta} = -\left(\frac{1}{\beta(\Delta t)^2}\frac{\partial \boldsymbol{M}}{\partial \theta} + \frac{\alpha}{\beta(\Delta t)}\frac{\partial \boldsymbol{C}}{\partial \theta}\right)\boldsymbol{u}_{n+1} - \frac{\partial \boldsymbol{R}(\boldsymbol{u}_{n+1}(\theta), \theta)}{\partial \theta}\Big|_{\boldsymbol{u}_{n+1}} + \frac{\partial \tilde{\boldsymbol{F}}_{n+1}}{\partial \theta} \tag{5}$$

here

$$\frac{\partial \tilde{\boldsymbol{F}}_{n+1}}{\partial \theta} = \frac{\partial \boldsymbol{F}_{n+1}}{\partial \theta} + \frac{\partial \boldsymbol{M}}{\partial \theta}\left(\frac{1}{\beta(\Delta t)^2}\boldsymbol{u}_n + \frac{1}{\beta(\Delta t)}\dot{\boldsymbol{u}}_n - \left(1 - \frac{1}{2\beta}\right)\ddot{\boldsymbol{u}}_n\right) + \boldsymbol{M}\left[\frac{1}{\beta(\Delta t)^2}\frac{\partial \boldsymbol{u}_n}{\partial \theta} + \frac{1}{\beta(\Delta t)}\frac{\partial \dot{\boldsymbol{u}}_n}{\partial \theta} - \left(1 - \frac{1}{2\beta}\right)\frac{\partial \ddot{\boldsymbol{u}}_n}{\partial \theta}\right] + \frac{\partial \boldsymbol{C}}{\partial \theta}\left[\frac{\alpha}{\beta(\Delta t)}\boldsymbol{u}_n - \left(1 - \frac{\alpha}{\beta}\right)\dot{\boldsymbol{u}}_n - (\Delta t)\left(1 - \frac{\alpha}{2\beta}\right)\ddot{\boldsymbol{u}}_n\right] + \boldsymbol{C}\left[\frac{\alpha}{\beta(\Delta t)}\frac{\partial \boldsymbol{u}_n}{\partial \theta} - \left(1 - \frac{\alpha}{\beta}\right)\frac{\partial \dot{\boldsymbol{u}}_n}{\partial \theta} - (\Delta t)\left(1 - \frac{\alpha}{2\beta}\right)\frac{\partial \ddot{\boldsymbol{u}}_n}{\partial \theta}\right]$$

In Eq. (5) the partial derivative of the internal resisting force vector $\boldsymbol{R}(\boldsymbol{u}_{n+1})$ with respect to sensitivity parameter θ can be presented by the term $[\partial \boldsymbol{R}(\boldsymbol{u}_{n+1}(\theta), \theta/\partial\theta]|_{\boldsymbol{u}_{n+1}}$ where the displacement vector $\boldsymbol{u}_{n+1}$ remains fixed.

In Eq. (5) $\boldsymbol{K}_T^{dyn} = 1/\beta(\Delta t)^2\boldsymbol{M} + \alpha/\beta(\Delta t)\boldsymbol{C} + \left(\boldsymbol{K}_T^{stat}\right)_{n+1}$ represents the tangent dynamic stiffness matrix, where, $\left(\boldsymbol{K}_T^{stat}\right)_{n+1}^{i}$ i is the *consistent tangent stiffness matrix* defined by $\left(\boldsymbol{K}_T^{stat}\right)_{n+1} = \partial \boldsymbol{R}_{n+1}/\partial \boldsymbol{u}_{n+1}$. The word 'consistent' emphasizes that the tangent operator is obtained through consistent linearization of the constitutive law integration scheme, which guarantees the quadratic rate of asymptotic convergence of the iterative solution strategies based on Newton's method (Simo and Taylor 1985).

4 FE STRUCTURAL MODEL

The present study consists of modeling and analysis of a RC frame (Fig. 4) subjected to earthquake loading using OpenSees. The structure is modeled using the displacement-based, Euler–Bernoulli frame element with the distributed plasticity (Gu et al. 2010). In each element five integration points are provided and section stress results are computed by discretizing the frame sections into layer. Two different constitutive concrete models are used to evaluate the response sensitivity of the material with the smoothing effect. The parameters used are: f_c = concrete compressive strength; f_u = concrete crushing strength; ε_0 = concrete strain at maximum strength; ε_u = concrete strain at crushing strength; E_c = initial tangent stiffness; and η = smoothing parameter. In the column section, different material parameters are used for the unconfined and confined concrete parts. In case of steel material, 1D J_2 plasticity and M–P models are considered. The steel parameters used as: E = Young's modulus; f_y = yield strength; H_{kin} = kinematic hardening modulus; H_{iso} = isotropic hardening modulus; b = strain-hardening ratio, R_0, cR_1, cR_2 = parameters which control the elastic to plastic transition; a_1, a_2 = isotropic hardening parameters. The input parametric values being used are listed in the Table 1 and 2. The dead and live loads are applied in the frame using lumped mass phenomenon at the nodal points.

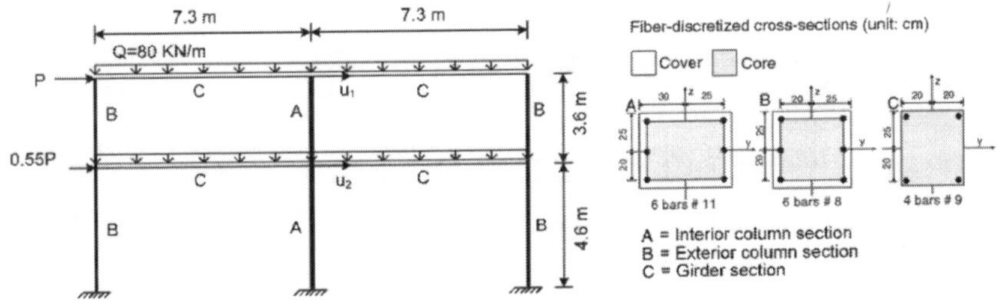

Fig. 4 Geometry, loading due to gravity and cross-sectional properties of frame structure.

Table 1 Parametric values of different concrete constitutive models.

Smoothed Popovics–Saenz concrete model			Kent–Scott–Park concrete model		
Parameter	Core concrete	Cover concrete	Parameter	Core concrete	Cover concrete
f_c (kPa)	34473.8	27579.04	f_c (kPa)	34485.6	27588.5
f_u (kPa)	25723.0	1000.0	f_u (kPa)	20691.4	0.0
ε_0	0.005	0.002	ε_0	0.004	0.002
ε_u	0.02	0.012	ε_u	0.014	0.008
E_c (kPa)	2.7851×10^7	2.4910×10^7			
η	0.2	0.2			

Table 2 Parametric values of different steel constitutive models.

M–P model		J_2 plasticity model	
E (kPa)	2.1×10^8	E (kPa)	2.1×10^8
f_y (kPa)	2.48×10^5	f_y (kPa)	2.48×10^5
b	0.02	H_{kin} (kPa)	1.6129×10^6
R_0	20	H_{iso}	0.0
cR_1	0.925		
cR_2	0.15		
a_1	18.5		
a_2	0.15		

The cyclic loading and dynamic base excitation are applied on the 2D frame to get response sensitivities. The lateral forces are applied at the corresponding nodes of each floor level as shown in Fig. 4. The total horizontal base acceleration used in the dynamic analysis (Fig. 5) is scaled by 10 with peak ground acceleration 14.89 m/s^2 (Zhang et al. 2008). Newmark-β method is used for the integration operations with parameters $\beta = 0.2756$ and $\gamma = 0.55$ with a constant time interval Δt=0.01[s] (Gu et al. 2009a). Response sensitivities are calculated for confined and unconfined concrete and reinforcing steel parameters by DDM as well as forward finite difference (FFD) method.

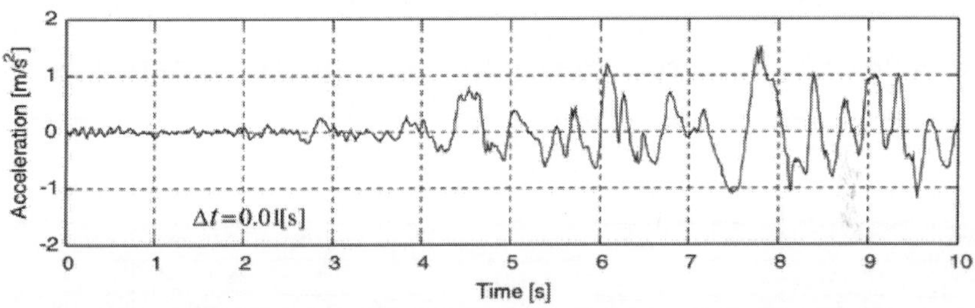

Fig. 5 Acceleration time history at the base of the RC frame.

FE reliability method is used to estimate probabilities of failure to achieve predefined performance. The performance functions are defined in terms of lateral displacement response of frame structure. For a simplest case of one performance function, the reliability problem can be defined as (Haukaas and Der Kiureghian 2004)

$$p_f = \int_{g(x) \leq 0} f(x)dx \tag{6}$$

where p_f is the failure probability, x is the FE model parameters vector, $g(x)$ is the performance function and $f(x)$ is the joint probability density function of x. An approximate solution of the above integral equation can be done by various available methods (e.g., FORM, sampling analysis). Advanced reliability method (FORM), and important sampling method are used to obtain failure probability and Hasofer–Lind (Hasofer and Lind 1974) reliability index β of frame structure. After the completion of the sensitivity analysis a static pushover study is carried out to evaluate the structural reliability performance. Applied lateral pushover load is same as the quasi-static cyclic loading. Smoothed Popovics–Saenz model has been chosen for further numerical study due to better performance in sensitivity analysis. All concrete and steel material parameters are characterized as random variables (RVs). All material RVs with lognormal distributions and probabilistic properties are listed in Table 3. Table 4 indicates the correlation coefficient between the RVs for each pair of members. The values of all RVs have been considered from another study (Gu 2008).

Table 3 Probabilistic properties of RC frame structure materials.

Concrete					Steel		
Material	Core		Cover		Material	Mean	Coefficient of variation
	Mean	Coefficient of variation	Mean	Coefficient of variation			
f_c (kPa)	34473.8	0.2	27579.04	0.2	E (kPa)	2.1 × 108	0.033
f_u (kPa)	25723.0	0.2	1000.0	–	f_y (kPa)	2.48 × 105	0.106
ε_0	0.005	0.2	0.002	0.2	b	0.02	0.2
ε_u	0.02	0.2	0.012	0.2			

Table 4 Correlation between the RVs.

RV	RV	Correlation
$f_{c(\text{core})}$	$f_{c(\text{cover})}$	0.8
$\varepsilon_{0(\text{core})}$	$\varepsilon_{0(\text{cover})}$	0.8
$\varepsilon_{u(\text{core})}$	$\varepsilon_{u(\text{cover})}$	0.8
$f_{c(\text{core})}$	$f_{u(\text{core})}$	0.8
$\varepsilon_{0(\text{core})}$	$\varepsilon_{u(\text{core})}$	0.8
$\varepsilon_{0(\text{cover})}$	$\varepsilon_{u(\text{cover})}$	0.8
$\varepsilon_{0(\text{core})}$	$\varepsilon_{u(\text{cover})}$	0.64
$f_{c(\text{core})}$	$\varepsilon_{0(\text{cover})}$	0.64
$f_{u(\text{core})}$	$f_{u(\text{cover})}$	0.64

Two different limit state functions (LSF) are considered:

$$g_1 = \Delta_{\text{lim}} - u_1 \tag{7}$$

$$g_2 = u_{\text{lim}} - (u_1 - u_2) \tag{8}$$

The first LSF (g_1) defines the failure event as the exceedance of the lateral roof displacement from 1 to 2 % of the building height, while the second LSF (g_2) defines a similar threshold for inter-story drift (Haukaas and Der Kiureghian 2004). Here, Δ_{lim} = 1–2 % of the building height, defines the first critical collapse displacement; u_{lim} = 1–2 % of the inter-story height is second critical collapse displacement; u_1 = roof displacement; and ($u_1 - u_2$) = inter-story drift. The freeware reliability tools called Risk tools (Rt) software is used to compute the structural reliability directly by using LSF equations (Mahsuli and Haukaas 2010). The OpenSees FE model is imported to Rt software, then all the RVs, their mean, standard deviation and correlation factors are inputted. Then, based on the failure functions (LSF), corresponding failure probability and reliability index are computed within Rt software framework.

5 RESULTS AND DISCUSSION

From the quasi-static cyclic analysis, the results are shown in Figs. 6 and 7. It is observed that, Popovics–Saenz concrete model with smoothing parameter (η) shows smooth continuous loading–unloading displacement response sensitivity curve under quasi-static cyclic loading (Fig. 6). On the other hand, the discontinuities in the response sensitivities are clearly visible from Popovics–Saenz concrete model without smoothing parameter and Kent–Scott–Park concrete model. The value of eta is used as 0.2 and obtained from the previous study (Gu et al.2010). Thus the smoothing parameter of Popovics–Saenz concrete model has a considerable contribution to achieve continuous response sensitivity in concrete constitutive model. For bi-linear steel (J_2 plasticity) model response sensitivities are discontinuous at the elastic-to-plastic transition points whereas, they are continuous throughout the curve for smooth steel model (Fig. 7). Previous findings by some researchers also indicate the same phenomenon in the similar manner (Haukaas and Der Kiureghian 2004; Barbato and Conte 2006). Thus, the continuity of the response computation can be achieved by using smooth material models which is essential for the convergence in the search of design point in reliability analysis.

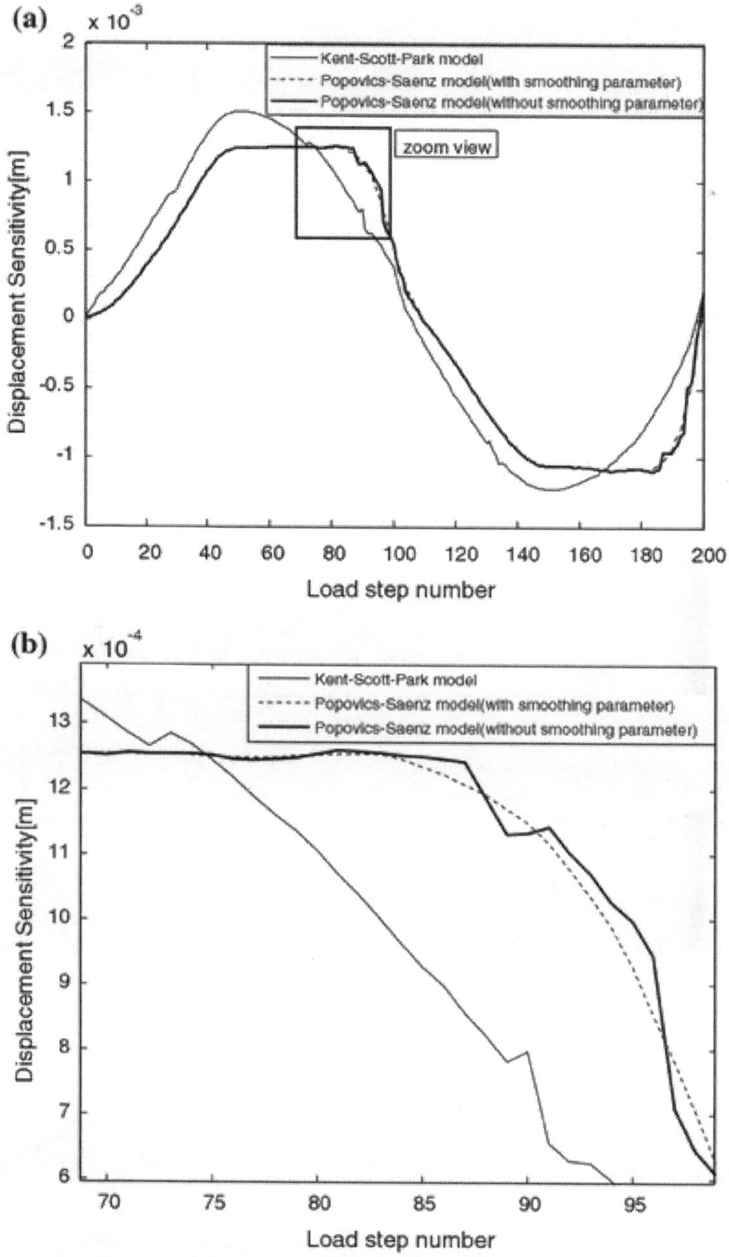

Fig. 6 **a** Displacement sensitivity with respect to core concrete compressive stress, f_c under static loading. **b** Zoom view.

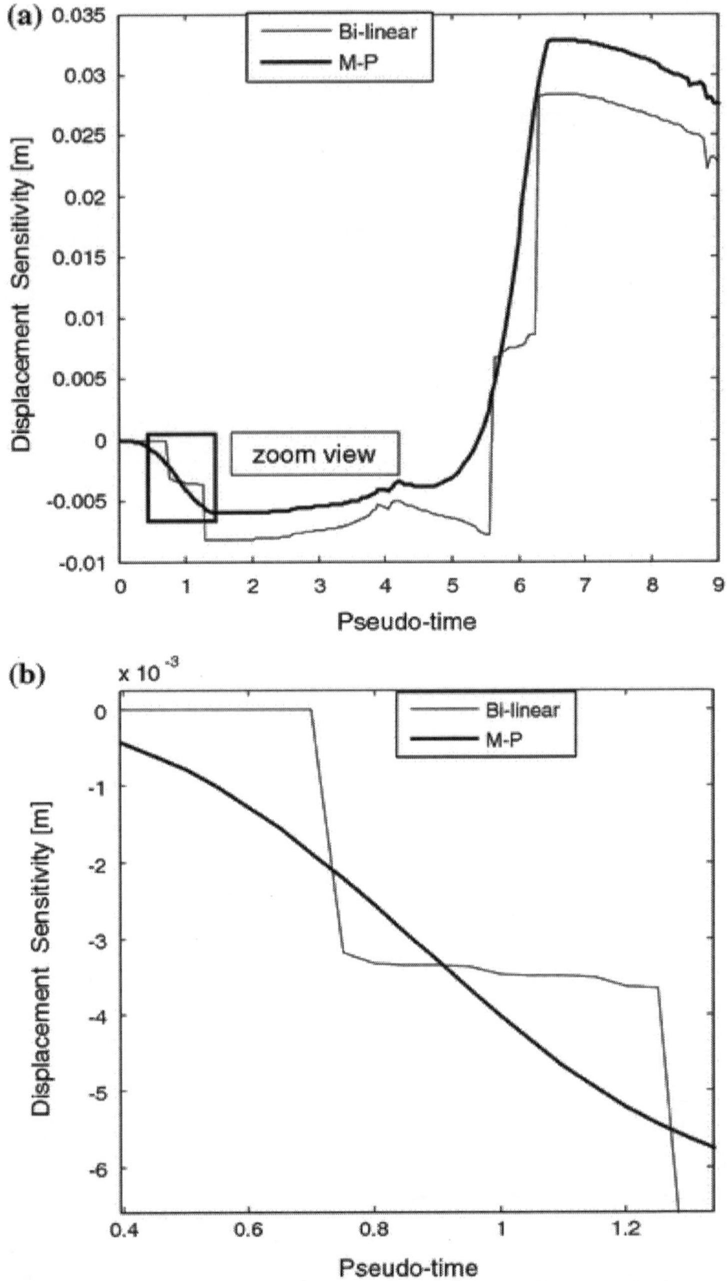

Fig. 7 a Displacement sensitivity with respect to steel Young's modulus, E under static loading.**b** Zoom view.

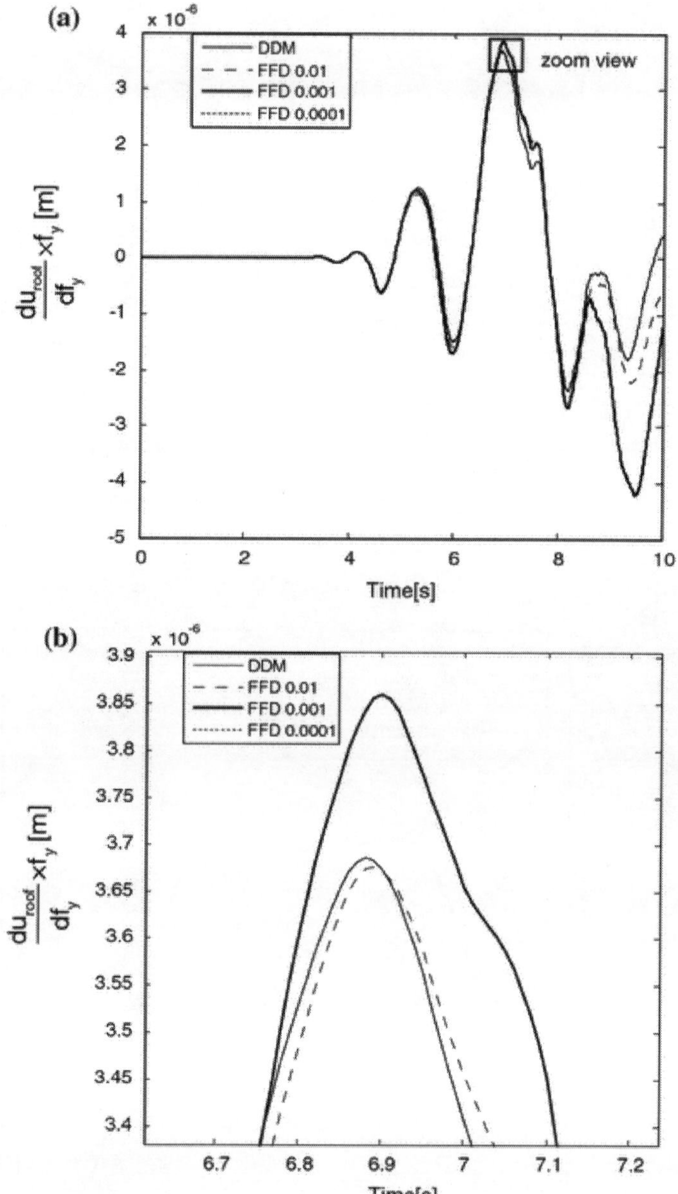

Fig. 8 **a** Validation of DDM results by FFD analysis: sensitivity of roof horizontal displacement response to the yield strength, f_y of steel. **b** Zoom view.

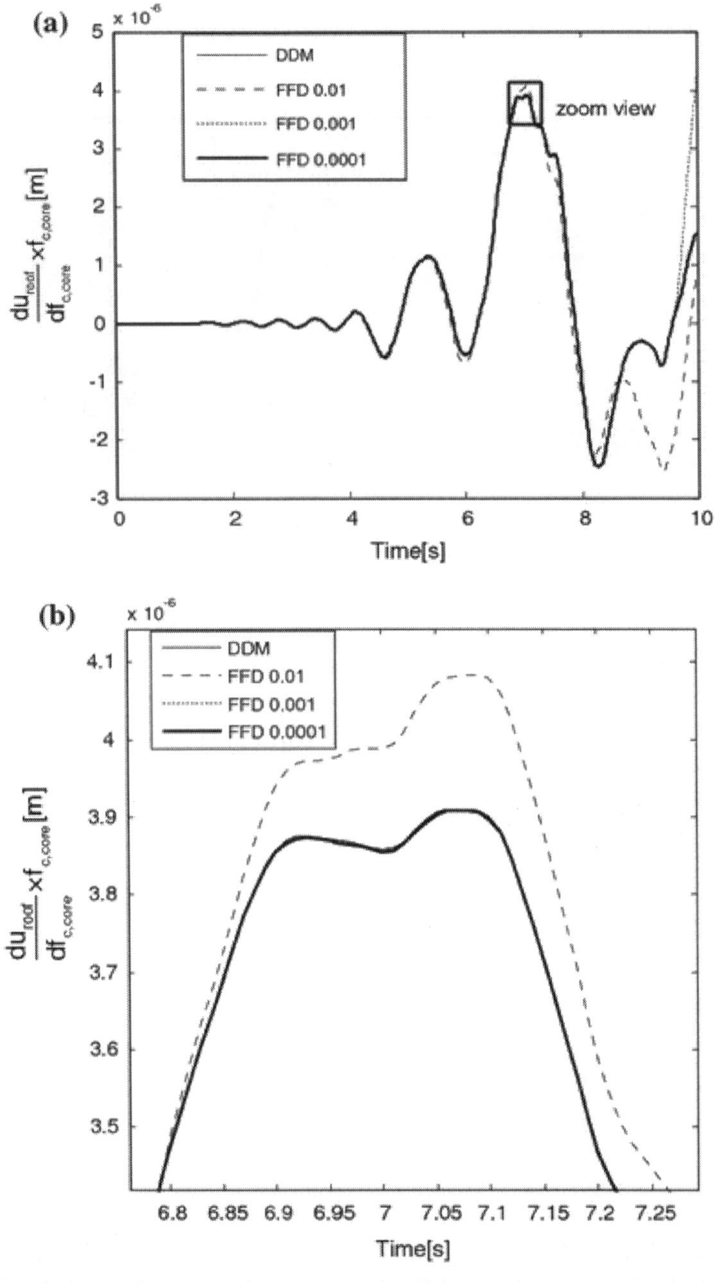

Fig. 9 **a** Validation of DDM results by FFD analysis: sensitivity of roof horizontal displacement response to the core concrete strength parameter, f_c. **b** Zoom view.

For the validation case, smoothed Popovics–Saenz concrete and M–P steel are used in 2D frame structure and earthquake motion is applied. Here the comparison between DDM and FFD method outcomes are shown for two material parameters. In first case, sensitivity of the lateral roof displacement due to the yield strength of reinforcing steel f_y is presented in Fig. 8. Three perturbation levels are considered, namely FFD 0 0.01, FFD 0.001, FFD 0.0001. Compressive strength of confined concrete, f_c is considered as the second parameter to validate the results (Fig. 9) and three levels of perturbation are considered as before. In both cases, FFD results converge well to the DDM results and the FFD results for perturbation level 0.0001 are close to the DDM results.

Figure 10 shows the displacement time history of the models used in this study. For the Kent–Scott–Park concrete model, maximum roof displacement to core concrete compressive strength f_c is more than that of smoothed Popovics–Saenz model. In both cases M–P reinforcing steel model has been used. It is inferred that under dynamic motions displacement responses of both concrete models are smooth along the time axis due to the inertia effects. Both of the models exhibit large plastic deformations under dynamic loading.

The structure is then modeled with M–P and J_2 plasticity reinforcing steel models. Figure 11demonstrates the computed time histories of the inter-story drift for two different steel models. Results corresponding to M–P model are not found very close to the ones obtained from the J_2plasticity model. It is also noticed that the maximum displacement occurs while using J_2plasticity model. Large plastic deformations are also observed at several points of constitutive steel models.

Figure 12 displays the sensitivity response of the lateral roof displacement with respect to the yield strength, f_y and steel Young's modulus, E under dynamic action. No discontinuities are found in the response sensitivities along the time axis for both steel models. Thus, the effect of inertia in the sensitivity equation is responsible to eliminate the discontinuities (Haukaas and Der Kiureghian 2004). Moreover, from the sensitivity results along the parameter axis, a different behavior is found in terms of discontinuity of the curves. Figure 12a, b shows small magnitude of discontinuities for response sensitivities obtained from the non-smooth J_2plasticity model, on the other hand M–P models are quite smoothen along the parameter axis. Table 5 shows the importance ranking of the material parameters. The normalized response sensitivity results for the inter-story drift $u_{drift} = u_1 - u_2$ are used to determine the significance of material parameters. Smoothed Popovics–Saenz concrete model and M–P steel model are used to select the effectiveness of material parameters. It is observed that parameter E, f_y, f_c are considered to be most important in earthquake excitations.

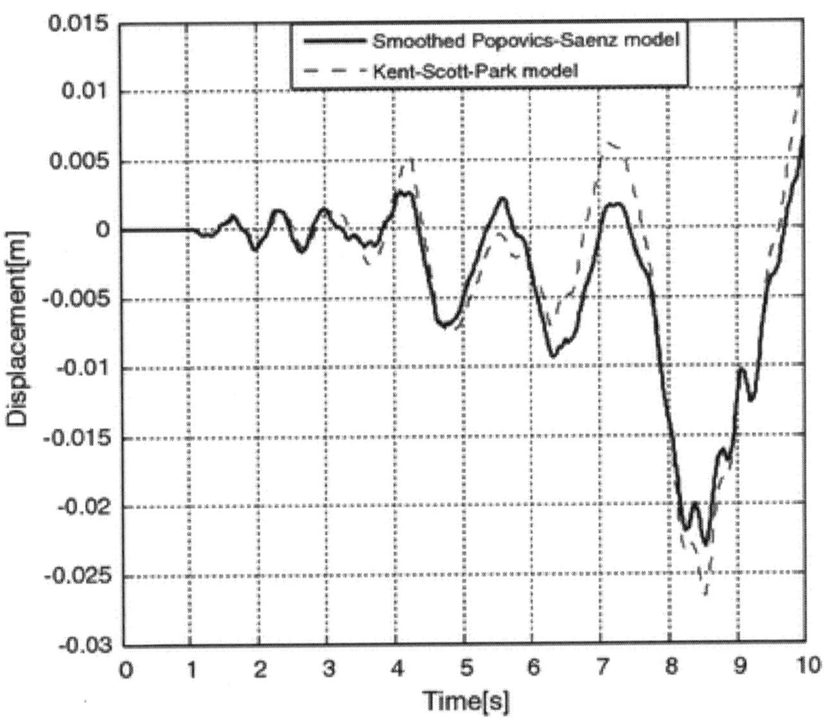

Fig. 10 Roof displacement responses due to core concrete compressive strength f_c of two different concrete models.

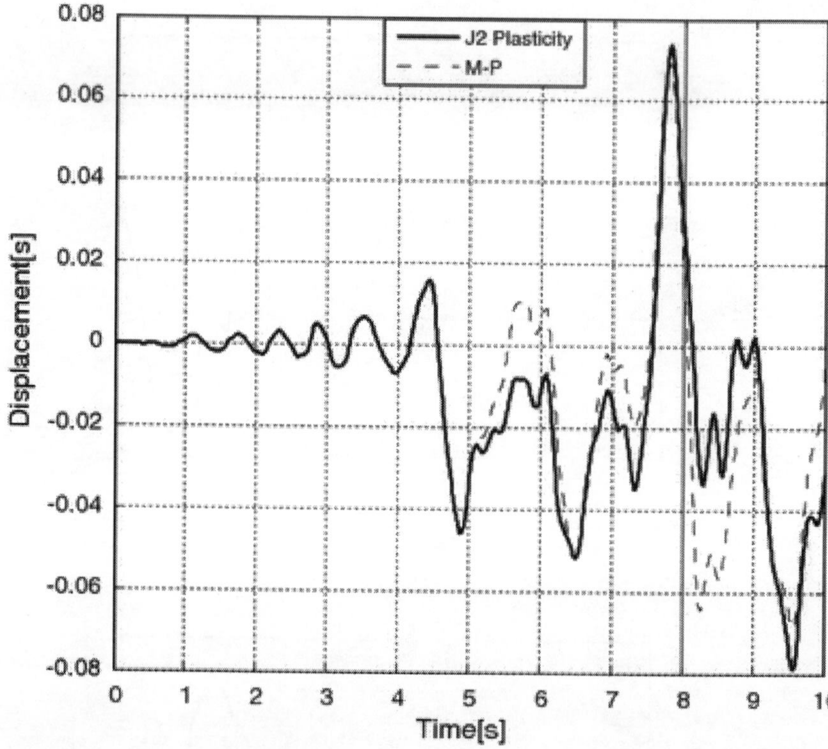

Fig. 11 Response histories of inter-story drift for different constitutive steel models.

Table 5 Parameter importance rankings for the inter-story drift of RC frame.

Response parameter	Concrete				Steel		
	Core		Cover				
	f_c	ε_u	f_c	ε_u	f_y	E	b
u_{drift}	4	5	3	6	2	1	7

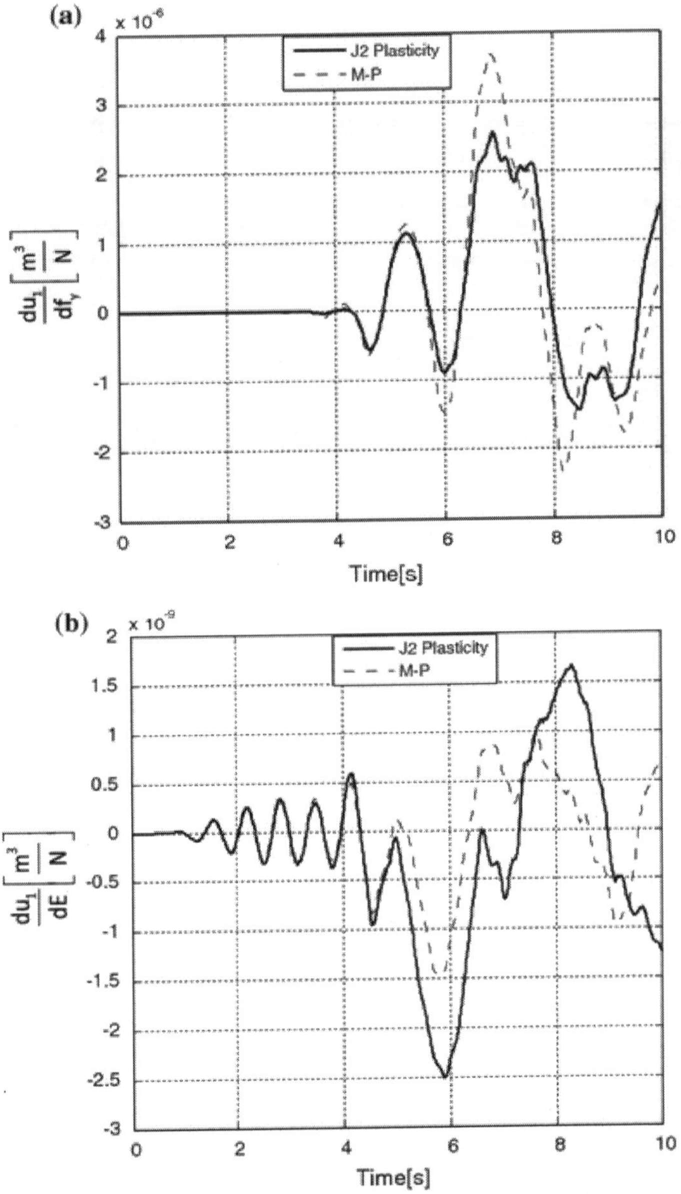

Fig. 12 a Sensitivity of roof horizontal displacement response with respect to f_y of steel. **b**Sensitivity of roof horizontal displacement response with respect to E of steel.

Reliability analysis results for the frame are also evaluated considering material parameters as RVs. Important sampling and FORM analyses are conducted for two LSF. Results in Table 6show good agreement between two methods which indicate the accuracy of FORM result in OpenSees framework. The outcomes also reflected that Reliability index β for LSF g_1 is more than LSF g_2. Thus, the structure is more reliable with respect to roof displacement rather than inter-story drift limit. Figures 13 and 14 present the CDF for the roof displacement and the inter-story drift of the frame using FORM analysis. At each point of the CDF curve one analysis run is required. The values of failure probabilities are obtained by changing the first and second critical collapse displacement values within a range of 1–2 % building height and inter-story height respectively, in Eqs. (7) and (8). The CDF curves of smooth (M–P) steel material are slightly different from non-smooth (J_2 plasticity) material in both cases. For 1.1 % of the roof displacement threshold, corresponding probabilities of failure for smoothed and non-smoothed material are 0.99976 and 0.986574 respectively. From CDF of the inter-story drift, very similar failure probabilities with that of the roof displacement are observed for 1.25 % of the inter-story drift threshold. Thus smoothing behavior of reinforcing steel has a minor effect on computed probabilities of the RC frame. Although the changes in reliability index are not noticeable, but an improvement in the search for design point has observed. It requires 182 s for reliability analysis with the bi-linear material, where it takes only 78 s for smooth (M–P) material. Therefore, the maximum trial point distance from the origin is much closer for smooth material than bi-linear material model. This advantage is achieved, because a large portion of the structure exhibits nonlinearity at the mean point by using smooth material model. Thus, using of smooth material model can be a fruitful technique to avoid non-convergence in FE based reliability analysis.

Table 6 Reliability index β for FORM and important sampling of RC frame.

LSF	FORM	Important sampling
g_1	3.19023	3.19118
g_2	3.02357	3.02583

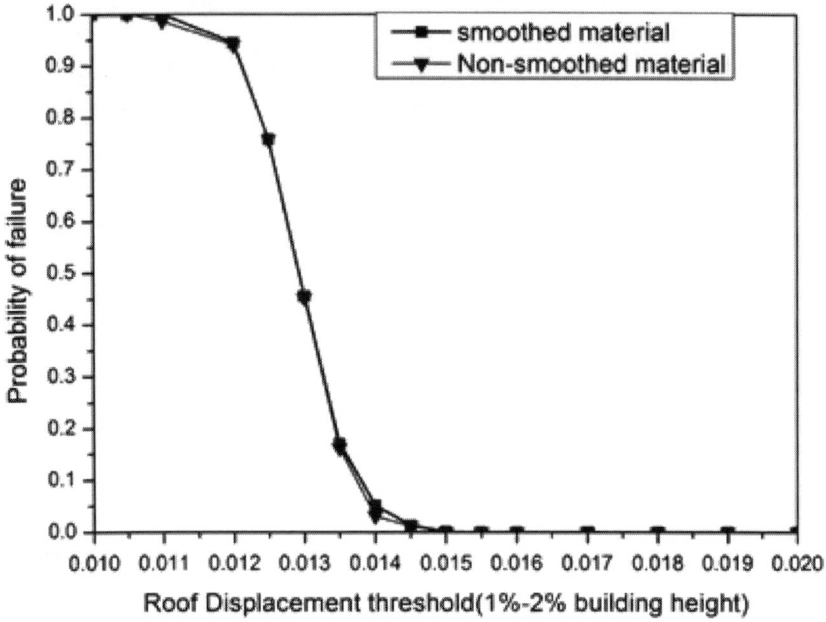

Fig. 13 CDF for roof displacement (threshold varies from 1 to 2 % of building height).

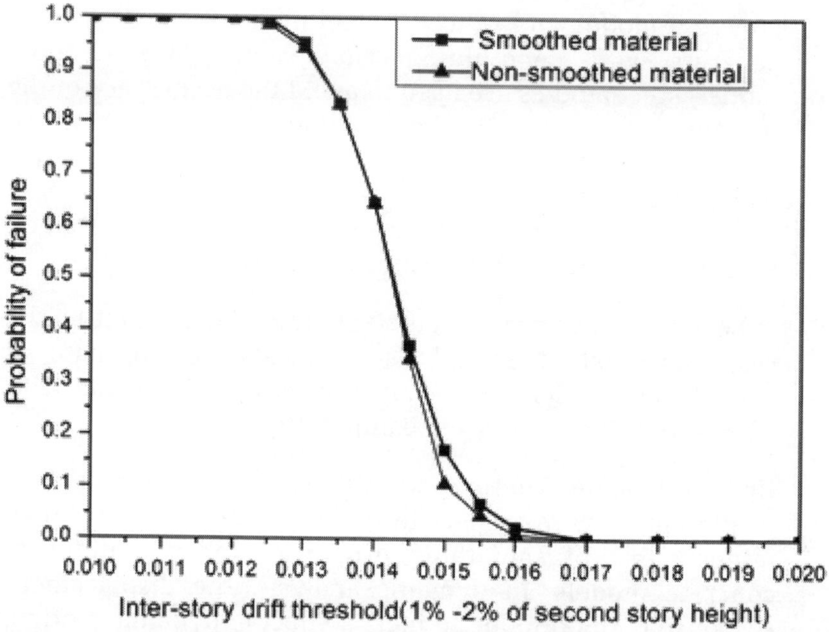

Fig. 14 CDF for inter-story drift (threshold varies from 1 to 2 % of second story height).

6 CONCLUSIONS

1. This work presents the analytical behavior of FE response sensitivities obtained from the different constitutive material models. DDM based nonlinear FE response sensitivity analysis is carried out under static and dynamic loading. FFD method has been used to investigate the accuracy of DDM. Smoothing behavior of both concrete and steel materials has been taken into consideration to evaluate the smoothness effects. After that, structural reliability analysis is performed to observe the smoothing effect of the reinforcing steel on failure probabilities of the frame. The core conclusions of this study can be summarized as follows:

2. In quasi-static FE analysis, response sensitivity discontinuities are observed in Popovics–Saenz concrete model without smoothing parameter, Kent–Scott–Park

concrete model, and bi-linear steel model. Popovics–Saenz concrete with smoothing parameter and M–P (smooth model) steel models are used to avoid the occurrence of the discontinuity and get the continuous computation of response sensitivity. Therefore, for the design point search, using gradient based optimization algorithms, smooth material model plays an important role to get accurate and efficient computations of response sensitivities.

3. DDM is very accurate and efficient method in computing the response sensitivities and also applicable to any material constitutive model. Good convergence of FFD results with DDM shows the accuracy in computation.

4. Under dynamic motion, sensitivity result of the roof displacement with respect to the core concrete has shown minor effect of smoothing properties for two different concrete models. In dynamic analysis, the displacement response is smooth due to inertia effect of dynamic motion. Linear inertia in the equation of motion also shows considerable smoothing effects on the response sensitivity results of steel constitutive models along the pseudo-time axis. Moreover, a minor discontinuity effect has been found in non-smooth J_2 plasticity steel along the parametric axis. This effect can be eliminated by using smooth steel model (M–P).

5. Structural reliability analysis shows the influence of smoothing on the computed probabilities of the RC structure. However, more efficient and reasonable reliability estimation can be achieved by using smooth material model compare with bilinear material constitutive model.

ACKNOWLEDGMENTS

This research was supported by the Ministry of Knowledge Economy and Korea Institute of Energy Technology Evaluation and

Planning (KETEP) as a part of the Nuclear R&D Program (No. 20101620100230). The authors would like to express their appreciation for the financial support.

REFERENCES

1. Arora, J. S., & Haug, E. J. (1979). Methods of design sensitivity analysis in structural optimization. *AIAA Journal, 17,* 970–974.

2. ATC-55. (2005). *Evaluation and improvement of inelastic seismic analysis procedures.* Advanced Technology Council, Redwood City, CA.

3. Balan, T. A., Filippou, F. C., & Popov, E. P. (1997). Constitutive model for 3D cyclic analysis of concrete structures. *Journal of Engineering Mechanics, 123*(2), 143–153.

4. Balan, T. A., Spacone, E., & Kwon, M. (2001). A 3D hypoplastic model for cyclic analysis of concrete structures. *Engineering Structures, 23,* 333–342.

5. Barbato, M., & Conte, J. P. (2006). Finite element structural response sensitivity and reliability analyses using smooth versus non-smooth material constitutive models. *International Journal of Reliability and Safety, 1*(1–2), 3–39.

6. BSSC. (2003). *The 2003 NEHRP recommended provisions for new buildings and other structures. Part 1: Provisions (FEMA 450).* Building Seismic Science Council.

7. Conte, J. P. (2001). Finite element response sensitivity analysis in earthquake engineering. In B. F. Spenser & Y. X. Hu (Eds.), *Earthquake engineering frontiers in the new millennium* (pp. 395–401). Lisse: Swets and Zeitlinger.

8. Conte, J. P., Barbato, M., & Spacone, E. (2004). Finite element response sensitivity analysis using force-based frame models. *International Journal of Numerical Methods in Engineering,59*(13), 1781–1820.

9. Conte, J. P., Vijalapura, P. K., & Meghella, M. (2003). Consistent finite-element response sensitivity analysis. *Journal of Engineering Mechanics, 129*(12), 1380–1393.

10. Der Kiureghian, A., & Ke, J. B. (1988). The stochastic finite element method in structural reliability. *Probabilistic Engineering Mechanics, 3*(2), 83–91.

11. Ditlevsen, O., & Madsen, H. O. (1996). *Structural reliability methods.* New York: Wiley.

12. Filippou, F. C., Popov, E. P., & Bertero, V. V. (1983). *Effects of bond deterioration on hysteretic behavior of reinforced concrete joints.* Report EERC 83-19, Earthquake Engineering Research Center, University of California, Berkeley, CA.

13. Gu, Q. (2008). *Finite element response sensitivity and reliability analysis of soil-foundation-structure-interaction systems.* Ph.D. dissertation, Department of Structural Engineering, University of California at San Diego, La Jolla, CA.

14. Gu, Q., Barbato, M., & Conte, J. P. (2009b). Handling of constraints in finite-element response sensitivity analysis. *Journal of Engineering Mechanics,135*(12), 1427–1438.

15. Gu, Q., Conte, J. P., & Barbato, M. (2010). *OpenSees command language manual response sensitivity analysis based on the direct differentiation method (DDM).* Berkeley, CA: Pacific Earthquake Engineering Center, University of California.

16. Gu, Q., Conte, J. P., Elgamal, A., & Yang, Z. (2009a). Finite element response sensitivity analysis of multi-yield-surface J2 plasticity model by direct differentiation method. *Computer Methods in Applied Mechanics and Engineering, 198*(30–32), 2272–2285.

17. Hasofer, A. M., & Lind, N. C. (1974). An exact and invariant first order reliability format. *Journal of Engineering Mechanics, 100*(12), 111–121.

18. Haukaas, T., & Der Kiureghian, A. (2004). *Finite element reliability and sensitivity methods for performance-based engineering*. Report PEER 2003/14, Pacific Earthquake Engineering Research Center, University of California, Berkeley, CA.

19. Hohenbichler, M., & Rackwitz, R. (1986). Sensitivity and importance measures in structural reliability. *Civil Engineering Systems, 3*(4), 203–209.

20. Kleiber, M., Antunez, H., Hien, T.D., & Kowalczyk, P. (1997). *Parameter sensitivity in nonlinear mechanics: Theory and finite element computations*. New York: Wiley.

21. Mahsuli, M., & Haukaas, T. (2010). Methods, models, and software for seismic risk analysis. In*9th US National & 10th Canadian Conference on Earthquake Engineering*, Toronto, 25–29 July 2010.

22. Mazzoni, S., McKenna, F., & Fenves, G. L. (2005). *OpenSees command language manual*. Berkeley, CA: Pacific Earthquake Engineering Center, University of California.

23. Menegotto, M., & Pinto, P. E. (1973). Method for analysis of cyclically loaded reinforced concrete plane frames including changes in geometry and non-elastic behavior of elements under combined normal force and bending. In *Proceedings, IABSE Symposium on'Resistance and Ultimate Deformability of Structures Acted on by Well-Defined Repeated Loads'*, Lisbon.

24. Monti, G., & Nuti, C. (1992). Non-linear cyclic behavior of reinforcing bars including buckling.*Journal of Structural Engineering, 118*(12), 3268–3284.

25. Schueller, G. I., Pradlwarter, H. J., & Koutsourelakis, P. S. (2004). A critical appraisal of reliability estimation procedures for high dimensions. *Probabilistic Engineering Mechanics,19*(4), 463–474.

26. Scott, M. H. (2012). Evaluation of force-based frame element response sensitivity formulations.*Journal of Structural Engineering, 138*(1), 72–80.

27. Scott, M. H., & Filippou, F. C. (2007). Response gradients for nonlinear beam-column elements under large displacements. *Journal of Structural Engineering, 133*(2), 155–165.

28. Scott, M. H., Franchin, P., Fenves, G. L., & Filippou, F. C. (2004). Response sensitivity for nonlinear beam-column elements. *Journal of Structural Engineering, 130*(9), 1281–1288.

29. Scott, B. D., Park, R., & Priestley, M. J. N. (1982). Stress–strain behavior of concrete confined by overlapping hoops at low and high strain rates. *ACI Journal, 79*(1), 13–27.

30. Simo, J. C., & Taylor, R. L. (1985). Consistent tangent operators for rate-independent elastoplasticity. *Computer Methods in Applied Mechanics and Engineering, 48,* 101–118.

31. Zhang,Y., Conte, J. P., Yang, Z., Elgamal, A., Bielak, J., & Acero, G. (2008). Two-dimensional nonlinear earthquake response analysis of a bridge-foundation-ground system. *Earthquake Spectra, 24*(2), 343–386.

32. Zhang, Y., & Der Kiureghian, A. (1993). Dynamic response sensitivity of inelastic structures. *Computer Methods in Applied Mechanics and Engineering, 108*(1–2), 23–36.

33. Zona, A., Barbato, M., & Conte, J. P. (2005). Finite element response sensitivity analysis of steel–concrete composite beams with deformable shear connection. *Journal of Engineering Mechanics, 131*(11), 1126–1139.

4

CHAPTER

Monitoring Instrumentation in Underground Structures

Alireza Maghsoudi, Behzad Kalantari

Department of Civil Engineering, University of Hormozgan, Bandar Abbas, Iran

ABSTRACT

Nowadays underground structures are very important. Based on observations of engineering; properties during geotechnical construction are an integral part of the design of underground structures. This research presents instrumentation as a tool to assist with these measurement observations, determine the need for modifications to loading or support arrangement. Also apart from above construction control, instrumentation is also indispensable for site investigation, design verification and safety of the structure. Instrumentation used in the construction of tunnels and subways can be implemented in three stages before, during operation and during operation are examined. Metro Railway Tunnels are constructed in populated area and have a more comprehensive instrumentation and monitoring program that additionally includes monitoring of ground conditions, underground water levels, tilt and settlement of nearby buildings or other structures of interest in the vicinity of the tunnel alignment. Instrumentation monitoring for metro railway tunnels includes monitoring of the structures under construction together with the ground, buildings and other facilities within the predicted zone of influence. Furthermore, instrumentation and subway tunnels in and around them increase accuracy of the different layers of the earth and excavation of the surrounding structures and make safety and accuracy. This paper presents the features of sophisticated instrumentation available today for geotechnical monitoring. A wide range of sophistic have been described with their applications ted electronic and mechanical instrumentation with different instrumentation schemes used to meet the requirements of different types of structures.

Keywords: Tunnel Hydraulic Fracturing, Flat Jack Test, Press Test (PMT), Plate Bearing, Test Extensometer, Piezometer, Seismic Refraction Tomography (SRT)

1. INTRODUCTION

Based on stability and strength of the subway tunnel and by instrument stability is a measure of confidence in the estimates. The initial cost of the used instruments, always look for a way to get the most information with the least utility. Deformation Measurements Instruments are installed in the tunnel roof and at selected points along the tunnel walls to monitor vertical, horizontal, and longitudinal in tunnel direction deformation components. The number of points and their detailed location depends on the size of the tunnel and the excavation sequencing in multiple drift applications. Also, as a minimum of the wall for each drift including temporary elements should be equipped with a device capable of measuring deformations [1] .

The designer of geotechnical construction works with naturally occurring materials, and does not find their exact engineering properties. He may carry out tests in the laboratory on the samples picked up from the field, and sometimes change the naturally occurring materials to make them more suitable for his needs. Instrumentation is used to measure the response (deformation, stress and etc.) of soil or rock to changes in loading or support arrangements, and from the measurements taken, the need for modifications to the loading or support arrangements is determined. This illustrates the basic reason why instrumentation is generally of immense value during geotechnical construction [2] .

2. FUNDAMENTALS OF INSTRUMENTATION

A good instrumentation program should have one or more of the following purposes in mind such as site investigation which instruments are used to characterize and determine initial site conditions. Common parameters of interest in a site investigation are pore pressure, permeability of soil, slope stability etc. In other side, design verification based on instruments are used to verify design assumptions. Instrumentation data from the initial stage of a project may show the need or provide the opportunity to modify the design in later stages. In this process Construction Control are instruments which installed to monitor the effects of construction. Instrument data helps the engineer to determine how fast construction can proceed without adverse effects on the foundation soil and construction materials used.

Safety of instruments can provide early warning of impending failure. In case of metro railway tunnels instruments provide early warning through real time monitoring systems available on the internet for any excessive and undue ground movements affecting the adjoining premises, structure and utilities like the railways, power lines, water lines etc. In this side, legal protection based on instruments provide designers and contractors the basis of a legal defence should resident and owners of adjacent properties blame construction for damage to their property and life. This aspect gains prominence in constructions in populated areas such as for underground metro railways.

Instrument performance is used to monitor the in-service performance of a structure. For example, monitoring leakage, pore water pressure and deformation can provide an indication of the performance of a dam. Monitoring loads on rock bolts and movements within a tunnel can provide an indication of the stability of tunnel [3] [4] .

3. TEST METHODS FOR UNDERGROUND STRUCTURES

3.1. In Situ Stress

In situ tests are used to directly obtain field measurements of useful soil and rock engineering properties. In soil, in situ testing include both index type tests, such as the Standard Penetration Test (SPT) and tests that determine the physical properties of the ground, such as shear strength from cone penetration Tests (CPT) and ground deformation properties from pressure meter tests (PMT). One significant property of interest in rock is it's in situ stress condition. Horizontal stresses of geological origin are often locked within the rock masses, resulted in a stress ratio (K) often higher than the number predicted by elastic theory.

3.2. Hydraulic Fracturing

Typically conducted in vertical boreholes. A short segment of the hole is sealed off using a straddle packer. This is followed by the pressurization by pumping in water. The pressure is raised until the rock surrounding the hole fails in tension at a critical pressure.

Flat Jack Test

Like all other stress measuring methods, the flat jack method is also based on certain assumptions such as 1) the stress at the test site is compressive, 2) the surrounding rock mass is homogeneous, elastic and isotropic, 3) the rock deforms symmetrically around the slot, 4) the state of stresses at the measurement site is uniform, and 5) the stress applied to the rock by the flat jack is uniform. The method can be used only in the walls of the underground opening after the opening has been made, and measures the induced stresses from which the in situ stresses are calculated. Flat jack, the Following breakdown, the shut-in pressure, and the lowest test-interval pressure at which the hydrocrack closes completely under the action of the stress acting normal to the hydro fractures. In a vertical test hole the hydro fractures are expected to be formed in vertical and perpendicular to the minimum horizontal stress)Figure 1(.

Some Common Mistakes form-meter, reference pins, a standard distance bar and a hydraulic pump with pressure gauges are the main items of equipment required for the test. Flat jack is inserted into the slot, cemented in place, and pressurized.

When the pins have been returned to the initial separation, the pressure in jack approximates the initial stress normal to the jack. The slot cut in the horizontal direction yields induced stress tangential to the boundary of the opening and the slot cut in the vertical direction yields induced stress, parallel to the axis of the opening, at the respective slot locations (Figure 2) [6] .

In actual tests, the length of the slot may be bigger than the jack, and the slot may not be of fixed width. Further, the stress acting in the plane parallel to the major axis of the slot also affects the contraction of the slot. The formulae given below account for the effect of all the above and can be used for evaluating the stresses (Equations (1) and (2)) [7] .

$PK_1 = P_\theta K_2 + P_H K_3$ (For jack position horizontal) (1)

$PK_1 = P_\theta K_3 + P_H K_2$ (For jack position vertical) (2)

4. MODULUS OF DEFORMATION

Modulus of deformation is recognized as one of the important parameters governing the rock mass behavior. It is experienced that tunnel sections generally take long time to stabilize. These time-dependent deformation effect the tunnel lining along with the modulus of deformation of the rock mass. The study suggests that time for ninety per cent tunnel closure and deformation is less both for good quality rock and larger span of tunnel. The concept of retarded creep appears to be valid for weak and dry rock masses around a supported tunnel in non-squeezing condition. The important various tests include [8] :

4.1. Pressure Meter Test (PMT)

Where P is the flat jack cancellation pressure, $P\theta$ is the induced stress tangential to the boundary of the opening and P_H is the induced stress parallel to the axis of the opening. K1, K2 and K3 are constants which are determined from parameters such as the length of the jack, slot dimensions, distance between reference pins, etc. Induced stresses Pθ and PH can be determined by solving simultaneous Equations (1) and (2). At each test site, several tests should be carried out to obtain statistically viable results. Typical stress-deformation plots from flat jack test data for very good quality basalt and poor quality volcanic breccia rock mass from Koyna H.E. Project, Maharashtra, with EM values of 62.8 GPa and 4.80 GPa, respectively, are shown in (Figure 3).

This test is performed with a cylindrical probe placed at the desired depth in a borehole. The Menard type pressure meter requires pre-drilling of the borehole; the self-boring type pressure meter advances the whole itself, thus reducing soil disturbance. The PENCEL pressure meter can be set in place by pressing it to the test depth or by direct driving from ground surface or from within a predrilled borehole. The hollow center PENCEL probe can be used in series with the static cone penetrometer.

The pressure required expanding the cylindrical membrane to a certain volume or radial strain is recorded in a borehole. It is applicable for soft rocks. Pressure meter are two types of High pressure instrument and Low pressure instrument (Figure 4) [9] .

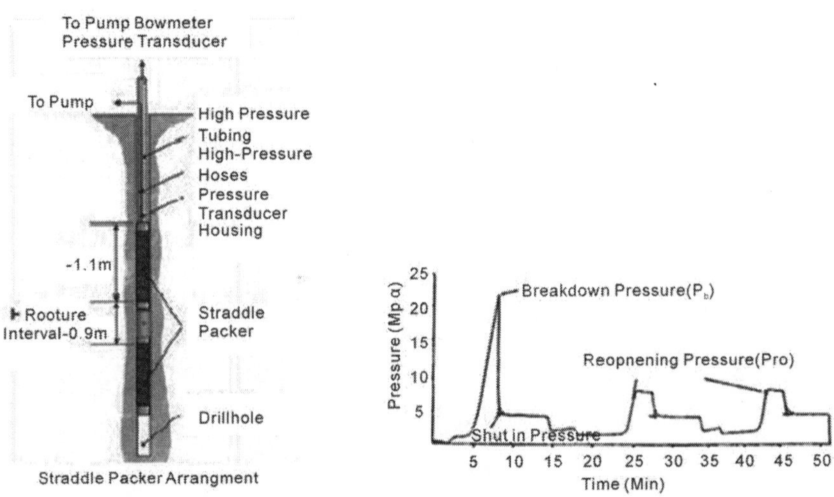

Figure 1. Schematic diagram and graphs of typical hydraulic fracturing system.

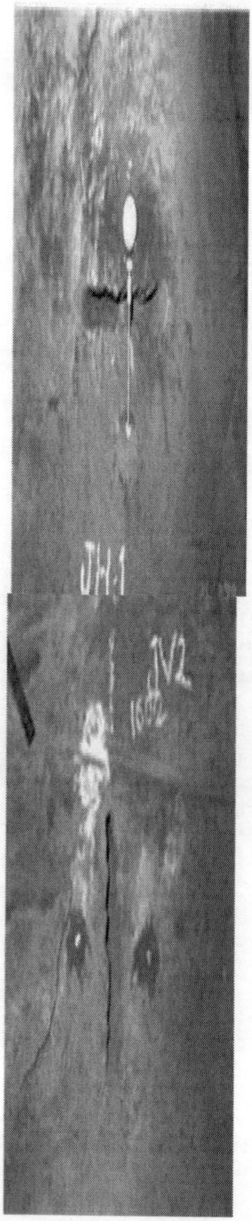
JH-1
JV2

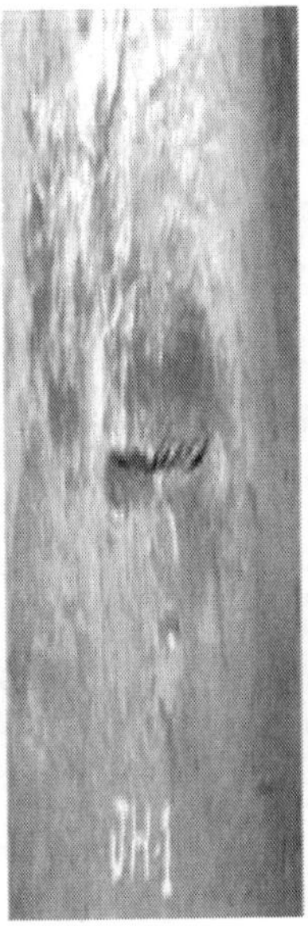

Figure 2. Slots made by jack hammer drilling and convergence measurement by deform-meter.

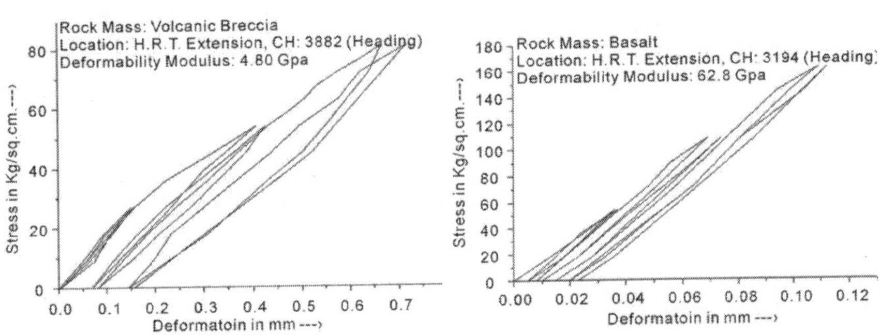

Figure 3. Typical stress verses deformation envelopes from flat jack tests.

The pressure meter test is an in-situ testing method which is commonly used to achieve a quick and easy measure of the in-situ stress-strain relationship of the soil which provides parameters such as the elastic modulus gives typical tests results (Figure 5). From the pressure strain curves, the untrained shear strength, the soil shear

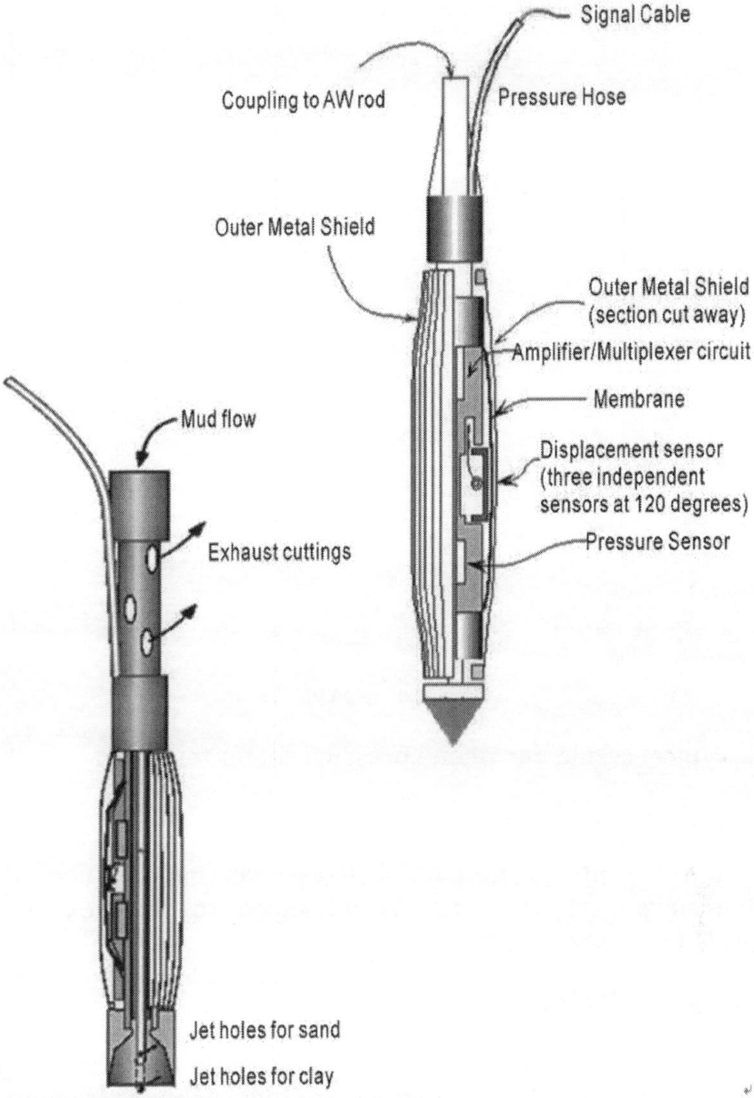

Figure 4. Low pressure instrument and high pressure instrument.

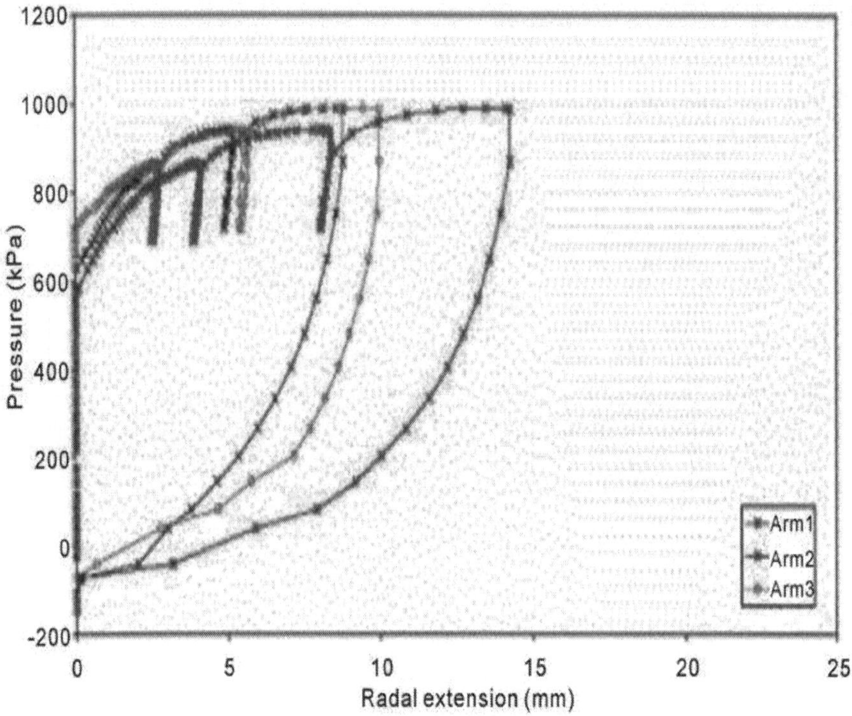

Figure 5. Typical results for a cone pressure meter test.

modulus and the in-situ horizontal stress can be obtained. On completion of a test, the cone is advanced to the next test depth [10] [11] .

4.2. Figures and Tables

In addition the pressure meter membrane can be inflated to a large radial strain, approximately 50%, to ensure that the pressure meter limit pressure is approached at full inflation. The speed of cone penetration testing enables the cone pressure meter to be an economical alternative to conventional self-boring and Menard-type pressure meter tests.

4.3. Plate Bearing Test

A relatively flat rock surface us sculptured and level with mortar to receive circular bearing plates 20 to 40 inches in diameter. Loading a rock surface and monitoring the resulting displacement. This is easily arranged in the underground gallery. The site may be selected carefully to exclude loose, highly fractured rock. The modulus of subgrade reaction is calculated from the relation [12] .

$$K = \frac{P}{0.125}\left(\frac{\text{Kg/cm}^2}{\text{cm}}\right)$$

5. MEASURING DISPLACEMENTS

The objectives of ground deformation monitoring are different in mountain and urban tunnels. In mountain tunnels, the main objective of deformation measurements during construction is to ensure that ground pressures are adequately controlled, i.e., there exists an adequate margin of safety against collapse, including roof collapse, bottom heave, failure of the excavation face, yielding of the support system, etc. Adequate control of ground pressures ensures a safe and economical structure, well adapted to the inherent heterogeneity of ground conditions. As a result of these differences in objectives, design philosophies, and construction techniques, the types and required accuracy of the measured ground deformations vary between the two classes of tunnels, as follows[13] :

One ft2 Plate is placed on a soil surfacee. A load P is placed on the plate and the settlementis measured as a function of P. Two different types of shapes can be developed depending on the density of the soil. From this, we can get a peak and residual bearing capacity. Notice the shape of the curves; they look a lot like the curves seen in a direct shear test. This is because the direct shear test reflects the conditions at failure (Figure 6).

1) In mountain tunnels, considerable ground deformations are deliberately permitted (and often provoked) in order to reduce the initially very large "geostatic" loads on the temporary support by increasing ground de-confinement.

2) In urban tunnels, the main objective is limiting ground deformations around the tunnel and thus causing the minimum

possible movement and disturbance at ground surface and the structures founded there.

The important various tests include extensometer. Extensometers can be installed in tunnel to measure the displacement of rock mass surrounding the tunnel. From the measured data of extensometers, some information of the rock mass, for example, the initial stresses and modulus of deformation, can be obtained. The displacements of extensometer are often induced by many inadequate factors, such as poor grouting at anchor point of extensometer [14] .

In a tunnel in Tehran is obtained using a extensometer. Station 10 is a full station in which convergence meter, extensometer and load cell installed that useful information can be obtained from this station such as: determination of plastic zone radius, determination of effective drilling zone around tunnel, stability and evaluation of tunnel using direct strain control technique.

Thus, some limitations of the measured data should be clarified before the measured data of extensometer are applied for further study. Figure 7 shows two types of instruments for measuring the vertical settlement of a series of locations in the ground.

Drawing of characteristic curve of the earth and determination of appropriate support system for tunnel. In figure 8 the actual rock mass displacement charts around the tunnel in the left and right walls are shown [15] .

6. DETERMINATION OF PORE PRESSURE

The soil is incorporated in the design of tunnels by calculating the soil pressures that are exerted on the tunnel and pressures necessary to have a stable tunnel face. Pore pressures are taken into account, but generally only a hydrostatic pressure distribution is assumed. However, the construction of a tunnel uses liquids pressurized with pressures different from the hydrostatic pressure: betonies slurry at the tunnel face and grout mortar at the

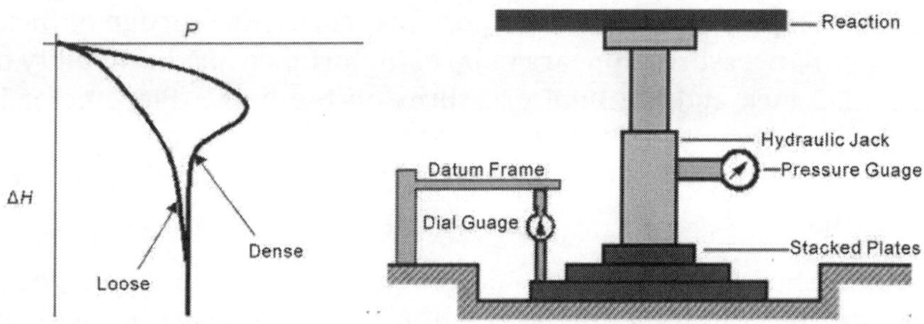

Figure 6. Plate bearing test (PBT) and plate bearing test results for loose and dense soils.

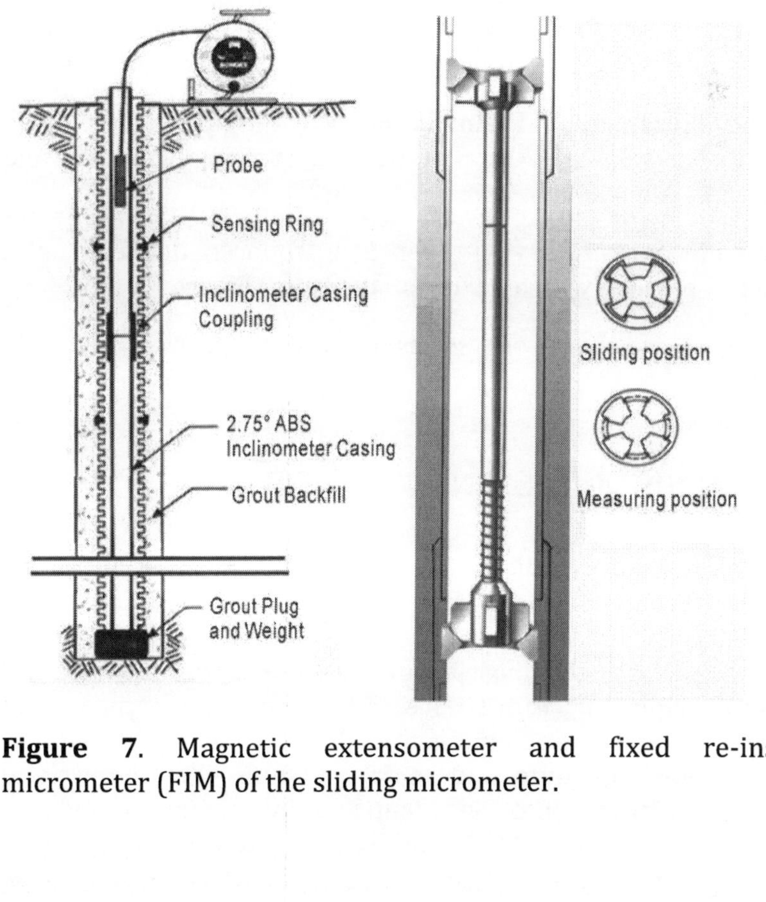

Figure 7. Magnetic extensometer and fixed re-installable micrometer (FIM) of the sliding micrometer.

tail void. This will lead to excess pore pressures. Knowledge of these excess pore pressures appears to be of importance for the stability of the tunnel face and the final pressures on the lining. Piezometer is the simplest form of Piezometric measurement is the monitoring of water levels in cased boreholes (internal diameter of casing 20 to 60 mm) [16] .

The plot below shows the results from a typical test in the height of the water level in the rising tube (piezometer) is measured with an electric water level gauge. The device is mounted on the end of the casing and the gauge body lowered into the tube. Once the electrode integrated in the plumb touches the surface of the water, an electric circuit is closed and a pilot lamp comes on. The depth is read off the scale on the cable. The height of the tube end is determined in advance by a levelling measurement. Pour water pressure measurement was carried out by various types of piezometers[17] :

Casagrande Piezometer: A simple open standpipe placed in a borehole, it consists of a PVC standpipe with a permeable membrane tip (Figure 9).

Pneumatic Piezometer: Dry nitrogen is introduced down an inlet tube which is blocked with a flexible diaphragm (Figure 9).

The first series and the second series of tests were run for 280 days (Figure 10).

7. GEOPHYSICAL TESTING METHODS

Geophysical tests are indirect methods of exploration in which changes in certain physical characteristics such as magnetism, density, electrical resistivity, elasticity, or a combination of these are used as an aid in developing subsurface information. Geophysical methods provide an expeditious and economical means of supplementing information obtained by direct exploratory methods, such as borings, test pits and in situ testing; identifying local anomalies that might not be identified by other methods of exploration; and defining strata boundaries between widely spaced borings for more realistic prediction of subsurface profiles. The important various tests include [18] :

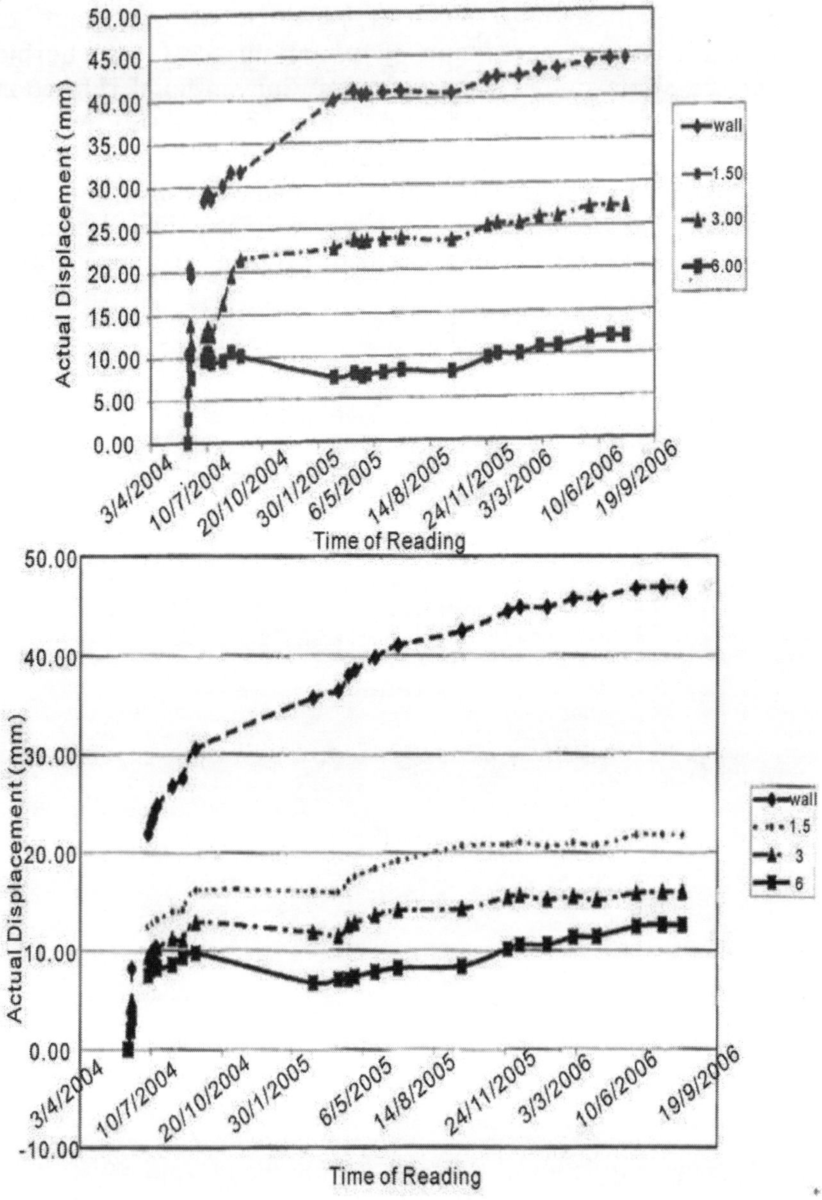

Figure 8. The actual displacement curve of the right rock mass and left rock mass in station 10 of Taloun pilot tunnel.

Seismic refraction tomography (SRT) which is important to investigate the loosing rock zone of tunnel surface for anchoring design of the some deformed rock mass. Conventional refraction inversion methods use a layered model approach. The subsurface is divided into a number of continuous constant velocity layers having velocities and thicknesses which are modified through interactive forward modeling in an effort to match travel times determined from the field data. These conventional methods require sections of travel time curves to be

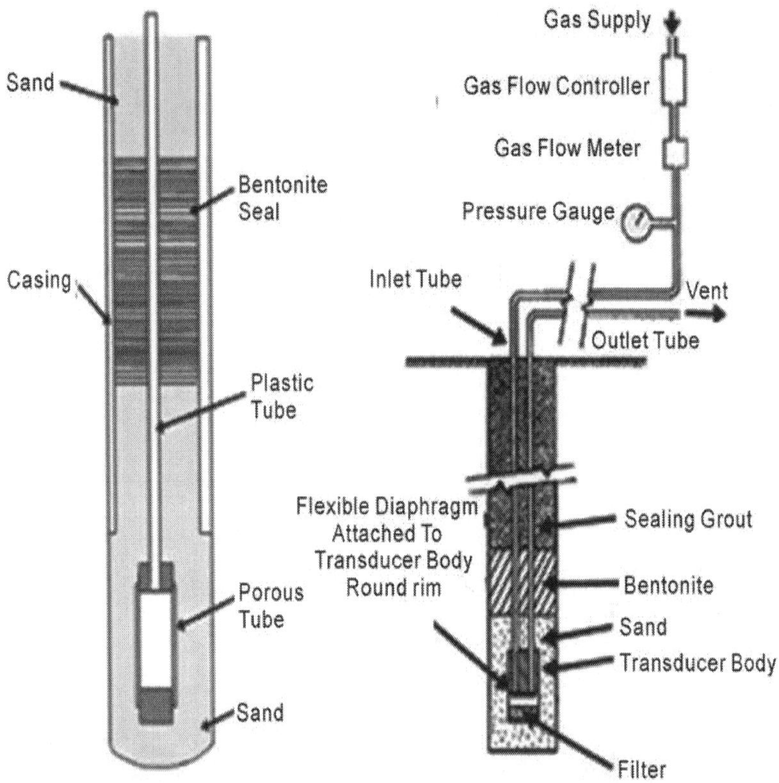

Figure 9. Casagrande Piezometer and Pneumatic Piezometer.

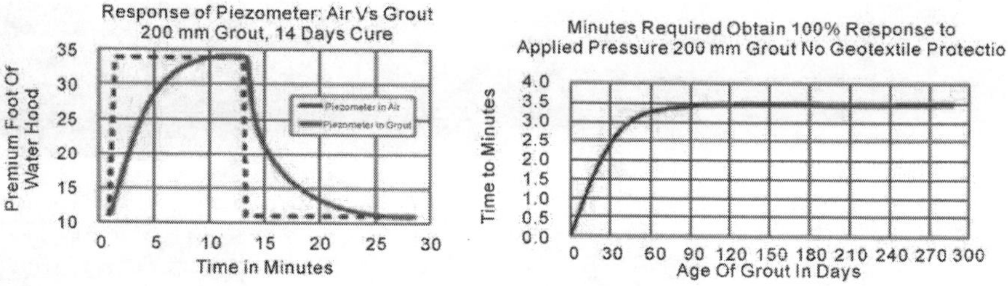

Figure 10. Casagrande piezometer and pneumatic piezometer.

mapped to refractors, a task that can be difficult in several geological situations. Detectors (geophones) are positioned on the ground surface at increasing distance from a seis mic impulse source, also at the ground surface. The time required for the seismic impulse to reach each geophone is recorded a seismic device like that shown in figure 11 shows contour lines that represented boundaries between compression wave velocity gradients [18] .

The cross-hole seismic technique was used to identify the extent of the Excavation Damage Zone (EDZ) is in a test tunnel and in a tunnel under construction. At least 2 boreholes are required: a source borehole within which a seismic pulse is generated, and a receiver borehole in which a geophone records generated compression and shear waves (figure 12). For increased accuracy additional receiver boreholes are used. Receivers must be properly oriented and securely in contact with the side of the borehole [19] . Boreholes deeper than about 30 ft should be surveyed using an inclinometer or other device to determine the travel distance between holes [20] , [21] .

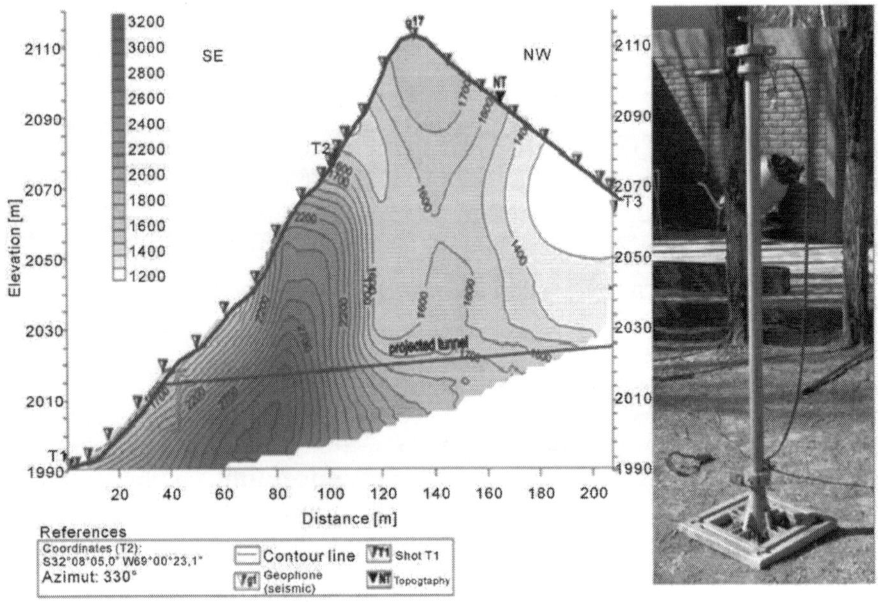

Figure 11. TAMPER seismic wave generation system and seismic refraction tomography profile.

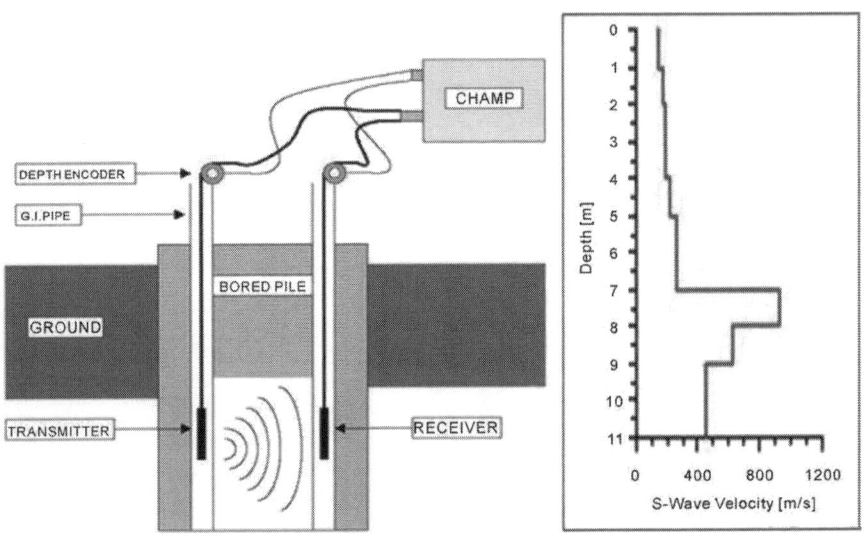

Figure 12. Cross-Hole devices and Resulting S-wave velocity profile.

8. CONCLUSION

Based on mentioned topic, laboratory testing is a fundamental element of a geotechnical investigation. The ultimate purpose of laboratory testing is to utilize repeatable procedures to refine the visual observations and field testing conducted as part of the subsurface field exploration program, and to determine how the soil or rock will behave under the imposed conditions. Laboratory testing of soil and rock samples to characterize materials; determine engineering properties for use in design; testing in accordance with specified standards; high-pressure 3-axial testing, large diameter 3-axial tests, dynamic testing with high-response, electro-hydraulic, closed loop MTS equipment, and high temperature testing; slake durability testing. Detailed soil laboratory testing is required to obtain accurate information including classification, characteristics, stiffness, strength, etc. for design and modeling purposes. Testing are performed on selected representative samples (disturbed and undisturbed) in accordance with ASTM standards. Rock Testing is generally controlled by the discontinuities within the rock mass and not the properties of the intact material. Standard rock testing evaluate physical properties of the rock included density and mineralogy in thin-section analysis. The mechanical properties of the intact rock core included uniaxial compressive strength, tensile strength, static and dynamic elastic constants, hardness of indices.

REFERENCES

1. Millogo, Y., Morel, J.-C., Traoré, K. and Ouedraogo R. (2012) Microstructure, Geotechnical and Mechanical Characteristics of Quicklime-Lateritic Gravels Mixtures Used in Road Construction. Construction and Building Materials, 26, 663-669.

2. Sikora, Z. and Ossowski, R. (2013) Geotechnical Aspects of Dike Construction Using Soil-Ash Composites. Procedia Engineering, 57, 1029-1035.

3. Argyroudis, S., Kaynia, A.M. and Pitilakis, K. (2013) Development of Fragility Functions for Geotechnical Constructions: Application to Cantilever Retaining Walls. Soil Dynamics and Earthquake Engineering, 50, 106-116.

4. Cabalar, A.F., Cevik, A. and Gokceoglu, C. (2012) Some Applications of Adaptive Neuro-Fuzzy Inference System (ANFIS) in Geotechnical Engineering. Computers and Geotechnics, 40, 14-33.

5. Kolat, C., Ulusay, R. and Suzen, M.L. (2012) Development of Geotechnical Microzonation Model for Yenisehir (Bursa, Turkey) Located at a Seismically Active Region. Engineering Geology, 127, 36-53.

6. Kim, H.-S., Cho, G.-C., Lee, J.Y. and Kim, S.-J. (2013) Geotechnical and Geophysical Properties of Deep Marine Fine-Grained Sediments Recovered during the Second Ulleung Basin Gas Hydrate Expedition, East Sea, Korea. Marine and Petroleum Geology, 47, 56-65.

7. Shaaban, F., Ismail, A., Massoud, U., Mesbah, H., Lethy, A. and Abbas, A.M. (2013) Geotechnical Assessment of Ground Conditions around a Tilted Building in Cairo-Egypt Using Geophysical Approaches. Journal of the Association of Arab Universities for Basic and Applied Sciences, 13, 63-72.

8. Jia, S.P., Zhao, Y.Q. and Zou, C.S. (2012) Numerical Solution to Identification Problems of Material Parameters in Geotechnical Engineering. Procedia Engineering, 28, 61-65.

9. Papaioannou, I. and Straub, D. (2012) Reliability Updating in Geotechnical Engineering Including Spatial Variability of Soil. Computers and Geotechnics, 42, 44-51.

10. Ghorbani, M., Sharifzadeh, M., Yasrobi, S. and Daiyan, M. (2012) Geotechnical, Structural and Geodetic Measurements for Conventional Tunnelling Hazards in Urban Areas—The Case of Niayesh Road Tunnel Project. Tunnelling and Underground Space Technology, 31, 1-8.

11. Al-Mukhtar, M., Khattab, S. and Alcover, J.-F. (2012) Microstructure and Geotechnical Properties of Lime-Treated Expansive Clayey Soil. Engineering Geology, 139-140, 17-27.

12. Ching, J.Y. and Phoon, K.-K. (2013) Quantile Value Method versus Design Value Method for Calibration of Reliability-Based Geotechnical Codes. Structural Safety, 44, 47-58.

13. Ghayoomi, M., Dashti, S. and McCartney, J.S. (2013) Performance of a Transparent Flexible Shear Beam Container for Geotechnical Centrifuge Modeling of Dynamic Problems. Soil Dynamics and Earthquake Engineering, 53, 230-239.

14. Juang, C.H. and Wang, L. (2013) Reliability-Based Robust Geotechnical Design of Spread Foundations Using Multi-Objective Genetic Algorithm. Computers and Geotechnics, 48, 96-106.

15. Ma, L., Luo, H.B. and Chen, H.R. (2013) Safety Risk Analysis Based on a Geotechnical Instrumentation Data Warehouse in Metro Tunnel Project. Automation in Construction, 34, 75-84.

16. Wu, X.Z. (2013) Trivariate Analysis of soil Ranking-Correlated Characteristics and Its Application to Probabilistic Stability Assessments in Geotechnical Engineering Problems. Soils and Foundations, 53, 540-556.

17. Katzenbach, R., Leppla, S., Vogler, M., Seip, M. and Kurze, S. (2013) Soil-Structure-Interaction of Tunnels and Superstructures during Construction and Service Time. Procedia Engineering, 57, 35-44.

18. Gaaver, K.E. (2012) Geotechnical Properties of Egyptian Collapsible Soils. Alexandria Engineering Journal, 51, 205-210.

19. Abela, J.M., Potts, D.M., Vollum, R.L. and Izzuddin, B.A. (2013) Geotechnical Analysis of Blinding Struts in Cut-And-Cover Excavations. Computers and Geotechnics, 48, 179-191.

20. Heerten, G. (2012) Reduction of Climate-Damaging Gases in Geotechnical Engineering Practice Using Geosynthetics. Geotextiles and Geomembranes, 30, 43-49.

21. Raptakis, D.G. (2012) Pre-Loading Effect on Dynamic Soil Properties: Seismic Methods and Their Efficiency in Geotechnical Aspects. Soil Dynamics and Earthquake Engineering, 34, 69-77.

5

CHAPTER

Seismic Reliability-Based Design Optimization of Reinforced Concrete Structures Including Soil-Structure Interaction Effects

Mohsen Khatibinia[1], Sadjad Gharehbaghi[2] and Abbas Moustafa[3]

[1] Department of Civil Engineering, University of Birjand, Birjand, Iran

[2] Department of Civil Engineering, Behbahan Khatam Alanbia University of Technology, Behbahan, Iran

[3] Department of Civil Engineering, Minia University, Minia, Egypt

1. INTRODUCTION

Past destructive earthquakes (e.g. the 1994 Northridge earthquake and the 1995 Kobe earthquake) have left a clear signature on the engineering community worldwide, changing thinking of structural engineers [1-2]. As such, after holding several workshops and conferences, an innovative approach namely Performance-Based Design (PBD) was presented by modern guidelines [3-5]. In principle, a structure designed using PBD approach should meet performance objectives in accordance with a set of specified reliabilities over its service life. This is aimed to reach structural design candidates associated with more predictable seismic behavior, quantifying and controlling the risk at an engineered acceptable level.

Both seismic demands and capacity parameters, that are inherently uncertain, are highly influential on the acceptable performance level of a structure. Furthermore, due to the fact that a structure on underlying soil is not rigid, soil-structure interaction (SSI) affects the responses of structures during an earthquake. Obviously, ignoring the SSI effects could lead to unrealistic structural responses and seismic demands. Hence, the effects of SSI should be considered in the seismic responses of structures [6]. Therefore, soil type, material properties of the structure, and ground motion characteristics randomly affect the seismic structural responses.

Deterministic structural optimization without considering the uncertainties in design manufacturing and operating processes may lead to unreliable design resulting in inappropriate balance between cost and safety. A proper design procedure must reasonably account for the inherent uncertain nature of a structural system and associated external load [7]. In structural optimization, non-deterministic performance of structures can be taken into account using robust design optimization (RDO) [8] and reliability-based design optimization (RBDO) [9]. RDO aims to minimize variation of the objective function, but RBDO optimizes the structural cost under reliability of the constraints.

A few studies have been implemented in a structural optimization problem, where RBDO is incorporated into PBD concept. Foley *et al.* [10] proposed a state-of-the-art model code and a PBD methodology. The methodology was applied to multiple-objective optimization problems for single storey and multi-storey structural frameworks with fully and partially restrained connections. Lagaros *et al.* [11] introduced a tool for the solution of realistic structural optimization problems that incorporate PBD under seismic loading into RBDO. The tool consisted of two distinctive methodologies based on artificial neural networks (ANNs). Fragiadakis *et al.* [12] presented optimum seismic design of reinforced (RC) structures considering the reliability constraints and PBD, where RBDO was implemented by the evolution strategies (ES). In the study by Moller *et al.* [13], the concept of PBD with RBDO was implemented by consideration of the uncertainties in the structural demands and capacities in order to evaluate reliability associated with each of the required performance levels. Khatibinia *et al.* [14] introduced RBDO of RC structures including SSI effects. In their study, the uncertainty of the structural demand was in terms of the random properties of the structure, underlying soil and the uncertain characteristics of artificially generated earthquakes. Also, the structural capacity associated with each of the required performance level in the concept of PBD was treated as an uncertain quantity.

Nonlinear dynamic analysis of structures using finite element method requires much computational effort. This drawback may accentuate when the nonlinear dynamic structural responses are required in RBDO of the structure using the Monte-Carlo Simulation (MCS) method, the importance sampling technique and the response surface method. In order to obtain an acceptable confidence within probabilities of the order close to 10^{-4} - 10^{-6}, the MCS method requires a large number of structural analyses. Based on Lagaros *et al.* [11], for such results, the large number of analyses ranges from 1.6×1061.6×106 to 1.6×1091.6×109 is required to achieve 95% likelihood for actual probability. To eliminate such drawbacks, utilizing soft computing-based models (e.g. artificial neural networks (ANNs)) is of crucial importance. These models can efficiently approximate structural responses and limit state functions in reliability analysis and optimization [11, 13, 15, 16]. Recently, as another model, support vector machines (SVMs) due to their simplicity, ease of implementation, and good performance have been developed to represent actual limit state functions. A hybrid of SVM and genetic algorithm, called (SVM-GA), was proposed to forecast reliability in engine systems [17]. The results showed the feasibility of SVM-GA in the reliability prediction compared with those of the ANNs and the autoregressive integrated moving average model. Zhiwei and Guangchen [18] investigated the capability of least squares support vector machine-based MCS (LSSVM-MCS) rather than SVM-MCS in reliability analysis. Based on the results of their study, LSSVM-MCS was more accurate and required less computational effort in comparison with SVM-MCS. Tan *et al.* [19] stipulated that there is no difference between the performance of the SVM-based response surface method (SVM-RSM) and the radial basis function neural network-based response surface method (RBFN-RSM). As a comparative study, Moura *et al.* [20] assessed the SVM effectiveness in forecasting time-to-failure and reliability of engineered components based on time series data. The efficacy of SVM with respect to other learning methods was shown. In the context of structural reliability assessment, a sampling method based on the adaptive Markov chain simulation and support vector density estimation was also developed by Dai *et al.* [21]. In their study, the application of SVM was proposed as a density estimator for structural reliability analysis.

In this chapter, RBDO of RC structures with considering SSI effects under time-history earthquake loading is presented in accordance with the PBD concept of SEAOC guidelines [3]. In this work, a new discrete gravitational search algorithm (DGSA) and an efficient proposed meta-model were introduced for performing RBDO of RC structures [22]. The objective function is the total cost of the structure while the constraints are treated as deterministic and probabilistic. The annual probability of non-performance for each performance level is considered as the probabilistic constraint in RBDO procedure. The new DGSA based on the fundamental concept of the standard GSA [23] is introduced for finding the optimal designs in the RBDO procedure. In DGSA, the position of each agent is presented in positive integer numbers. Also, the velocity of each agent is modified based on the particle swarm optimizer with passive congregation (PSOPC) which was proposed by He *et al.* [24]. The modifications can improve the global exploration ability of DGSA and overcome the shortcomings of the Binary GSA (BGSA) model introduced by Rashedi *et al.* [25]. A meta-model-based MCS method is also presented herein to take into account the probabilistic constraint in conjunction with the nonlinear finite element analysis (FEM) of SSI system. Due to the fact that the computational cost of MCS for structural reliability analysis is high, the meta-model is proposed to predict the structural seismic responses of SSI system, and significantly reduce the computational effort of the RBDO process. The meta-model is a combination of weighted least squares support vector machine (WLS-SVM) [26] and Morlet wavelet kernel function, which is called WWLS-SVM [14, 22, 27]. The selection of WWLS-SVM parameters efficiently affects the prediction accuracy of WWLS-SVM. Hence, the parameters of WWLS-SVM and wavelet kernel are assigned by using the standard gravitational search algorithm (GSA).

Numerical examples show that the wavelet as a kernel function is much better than those of the common kinds as kernel function in WLS-SVM. The accuracy and generalization of WWLS-SVM is improved using GSA. Furthermore, numerical results demonstrate the efficiency and computational advantages of the proposed DGSA for RBDO of structures.

2. RBDO OF RC STRUCTURES

2.1. FORMULATION OF OPTIMIZATION

Seismic design optimization of RC structures under time-history earthquake loads is an ongoing research topic and has received great attention among researchers [14, 27-33]. As such, RBDO of RC structures with the consideration of SSI effects was investigated in accordance with PBD concept of SEAOC guidelines [3] under seismic loading. This work incorporates the acceptable performance levels and the RBDO theory to compare the achieved annual probability of non-performance with target values for each performance level. The objective of the RBDO problem is to minimize the total cost whereas the deterministic and probabilistic constraints should not exceed a specified target.

The RBDO problem of RC structures can be formulated in the following form:

$$\begin{aligned}
&\text{Minimize} \quad C_{TOT} \\
&\text{Subject to} \quad g_i\left(\mathbf{X},\bar{\mathbf{X}}_{rand}\right) \le 0 \qquad i=1,2,\dots,N_g \\
&\qquad \beta_{annual}\left[G_j\left(\mathbf{X},\bar{\mathbf{X}}_{rand}\right) \ge 0\right] \ge \beta_j^{Target} \qquad j=1,2,3 \\
&\qquad \mathbf{X}=\{x_1,x_2,\dots,x_k,\dots,x_n\} \in R^d
\end{aligned}$$

where *CTOT* is the total cost; βannual is the system reliability index corresponding to the *j*th performance level (performance function), $G_j(\mathbf{X},\ \bar{\mathbf{X}}_{rand})$; β_j^{Target} is the prescribed target value of the reliability index; *Ng* is the number of deterministic constraints, gi(X)gi(X) ; A given set of discrete values,***X***, is expressed by R^d and design variables, *xk,* can take values only from this set. The vector $\bar{\mathbf{X}}_{rand}$ represents the random variables.

2.2. LIFE- CYCLE COST ASSESSMENT OF RC STRUCTURE

The total cost, *CTOT,* of a structure is the initial structural cost for a new structure construction and the repair cost from an earthquake and different levels of damage that may occur during the life of structure. This cost can be expressed as a function of the design vector ***X*** and the time *t* as follows:

$$C_{TOT}\left(\mathbf{X},\bar{\mathbf{X}}_{rand},t\right)=C_{IC}\left(\mathbf{X},\bar{\mathbf{X}}_{rand}\right)+C_{RC}\left(\mathbf{X},\bar{\mathbf{X}}_{rand},t\right)$$

where *CIC* is the initial cost of structure; *CRC* is the present value of the repair cost. In this study, the initial cost is considered as the sum of the total cost of concrete, *CC,* and the total cost of reinforcing steel, *CS,* is given [14]:

$$C_{IC}=C_C+C_S=\sum_{i-1}^{Ne} w_C b_i h_i L_i+\sum_{i-1}^{Ne} w_S A_{Si} L_i$$

where wC and wS are the unit cost coefficients of each material; *bi, hi,* and *Li* are the section dimensions of *i*th element and its length; and *ASi* is the total reinforcement in the section of element. *Ne*is the number of the elements of the structure.

The repair cost refers to the cost of damage level from earthquake that may occur during the life of a structure. In this study, the overall damage index, *DIoverall,* is considered as an indicator of structural damage. The total expected cost of repair based on the overall damage index is expressed as follows [13]:

$$C_{RC}=\int_0^1 C_{RC}\big|_{DI_{overall}} \cdot f_{DI_{overall}}\left(DI_{overall}\right)\cdot d\left(DI_{overall}\right)$$

where $f_{DI_{overall}}$ is the probability density function for the index, $DI_{overall}$. $C_{RC}\big|_{DI_{overall}}$

is the expected present cost which is defined as [34]:

$$C_{RC}\big|_{DI_{overall}}=\int_0^{\infty} C_{f0}\left(DI_{overall}\right)f\left(t\right)dt=C_f\left(DI_{overall}\right)\frac{\upsilon}{r+\upsilon}$$

$$C_f\left(DI_{overall}\right)=\begin{cases} kC_0\left(\dfrac{DI_{overall}}{0.60}\right) & \text{if } DI_{overall}\leq 0.60 \\ kC_0 & \text{if } DI_{overall}>0.60 \end{cases}$$

where υ and *r* are the mean occurrence of earthquakes and the discount rate, respectively. *C0* is the complete replacement cost, with k=1.20 assumed to be a factor to account for demolition and clearing. In this study, the value of 0.04 is assumed for the discount rate, *r*.

2.3. CONSTRAINT HANDLING APPROACH

A comprehensive overview of the most popular constraint handling approaches used in conjunction with meta-heuristic optimization methods was presented in the literature review by Coello Coello [35]. In the present study, the external penalty function method as one of the most common forms of the penalty function in the structural optimization [15, 27-29, 36-39] is employed to transform constrained RBDO problem into unconstrained one as follows:

$$fit\left(\mathbf{X},\bar{\mathbf{X}}_{rand}\right)=C_{TOT}\left(1+r_p PF\right)$$

where fit(X), *PF* and *rp* are the modified function, the penalty function, and an adjusting coefficient, respectively. The penalty function based on the violation of normalized constraints [36] is defined as the sum of all active constraints violations as indicated:

$$PF=\sum_i\left[\max\left(g_i,0\right)\right]^2+\sum_j\left[\max\left(1-\frac{\beta_{annual,j}}{\beta_j^{Target}},0\right)\right]^2$$

This formulation allows solutions with violated constraints, and the objective function is always greater than the non-violated one.

3. RELIABILITY ASSESSMENT OF RC STRUCTURE

In order to evaluate the system reliability index corresponding to each of the performance levels, RC structures should be assessed in the RBDO procedure [14, 22]. The system reliability index corresponding to each of the performance levels are estimated by MCS method. In the following subsections, the procedure of assessment of RC structures is explained.

3.1. REQUIRED DATABASE

In PBD approach, many uncertain variables influence the structural seismic responses. In the studies by Khatibinia *et al.* [14, 22], material properties of concrete, steel and soil, as well as earthquakes are considered as intervening uncertain variables. Because of a few historical records of earthquake for a selected site, selection of a proper ground motion record for a site is often difficult, even impossible in some cases. To overcome this problem, artificial earthquakes, statistically influenced by desired properties of a selected site, are utilized in seismic design of structures. The spectral representation method based on time domain procedure can be used for generation of artificial earthquakes [40]. The proper parameters for generation of artificial earthquakes are selected according by values proposed by Möller et al. [13]. In order to perform RBDO of RC structures using the proposed meta-model-based MCS, samples are generated randomly, and are used to train and test the meta-model. Inputs of the meta-model include the random combinations of intervening variables. In order to generate the database, seven random Peak Ground Acceleration (PGA) values are chosen. The PGA values are equal to 260, 350, 400, 550, 650, 700 and 800 (cm/sec^2). Accordingly, by using the Latin Hypercube Design (LHD) sampling method [41], and considering the 120 combinations based on the intervening variables for each PGA value, the total number 840 combinations are generated. Then, corresponding to the each PGA of each 840 combinations, five artificial earthquakes, as sub-combinations, with random phase angles are generated corresponding to each PGA value. For each of the 840 combinations, in the first phase, the structure is analyzed subjected to the combination of gravity loads according to ACI code [42]. Steel reinforcement ratios of longitudinal bars of structural elements' cross-sections, ρ, shall be satisfied in accordance with ACI code [42]. Furthermore, when choosing the steel reinforcement ratios, it is verified that they are sufficient to provide adequate strength against the combination of gravity loads. Based on ACI code [42], the strong column-weak beam concept shall be satisfied for every joint of designed structures for earthquake loads. In the second phase, nonlinear dynamic analysis of SSI system is performed for each sub-combination and seismic responses of SSI system are obtained. Using nonlinear dynamic analysis of SSI system, maximum roof displacement, *umax*, maximum inter-storey drift, *DRmax*, maximum local damage index, *DILmax*, and overall damage index, *DIoverall*, are considered as the structural seismic responses. After that, the mean, $\bar{R}_{i}$, and the standard deviation, σRi, of *i*th seismic response, *Ri*, corresponding to each of the 840 combinations subjected to five artificial earthquakes, are achieved. In this work, the modified Park-Ang damage index [43] as one of the most acceptable indices in seismic damage analysis of structures is used for calculating *DILmax* and*DIoverall.*

3.2. LIMIT STATE FUNCTIONS

The operational, life safety and collapse prevention levels have been defined as the performance levels. A performance level depends on some limit state functions. A limit state function, $G(\bar{X})$, is determined by capacity, R_{LIM}, and demand, $R(\bar{X})$, as follows [13, 22]:

$$G(\bar{X}) = R_{LIM}(\mu, \delta) - R(\bar{X})$$

where RLIM is the limiting value for a seismic response $R(\bar{X})$ at a given performance level, with mean value of μ and coefficient of variation δ.

The limit state functions and their probability distribution function (PDF) for the performance levels, according to SEAOC guidelines (2000), are shown in Table 1.

TABLE 1. The limit sate functions of the performance levels [14, 22]

Performance level	Limit sate function	$G(\bar{X})$	PDF
Operational			
	Elastic roof displacement	$(\bar{u}_y,\ 0.1)-u_{max}(\bar{X})$	Lognormal
	Inter-storey drift	$(0.005,\ 0.1)-DR_{max}(\bar{X})$	Lognormal
Life safety			
	Inter-storey drift	$(0.015,\ 0.1)-DR_{max}(\bar{X})$	Lognormal
	Max. local damage index	$(0.6,\ 0.1)-DIL_{max}(\bar{X})$	Beta
	Global damage index	$(0.4,\ 0.1)-DI_{overall}(\bar{X})$	Beta
Collapse			
	Inter-storey drift	$(0.025,\ 0.1)-DR_{max}(\bar{X})$	Lognormal
	Max. local damage index	$(1,\ 0.1)-DIL_{max}(\bar{X})$	Beta
	Global damage index	$(0.4,\ 0.1)-DI_{overall}(\bar{X})$	Beta

In Table 1, $\bar{u}_y$ is the yielding horizontal displacement at top storey of the frame which is determined by a pushover analysis. Furthermore, the demand, $R(\bar{X})$, corresponding to each of seismic responses are defined using the mean, $\bar{R}$, and the standard deviation, σR.

3.3. ANNUAL PROBABILITY OF NON-PERFORMANCE

The non-performance probability, Pf, is considered as a function of the limit state functions in proportion to a specified performance level. Using the evaluation of the multiple integral over the failure domain, $G(\bar{X})\leq 0$, P_f

is calculated as follows:

$$P_f = \int_{G(\bar{X})\leq 0} \int \cdots \int f_{\bar{X}}\left(\bar{X}\right) d\bar{X}$$

where $f_{\bar{X}}(\bar{X})$ is the joint probability density function of $\bar{X}$.

For each performance level, the total exceeding probability, PfE, is considered as a series system. Determination of the total exceeding probability, PfE, is based on integration of a multi-normal distribution function. In order to estimate the integral, the MCS method is used concurrently for all limit state functions in proportional to the performance levels listed in Table 1. Therefore, the seismic reliability corresponding to the each performance level is defined by an annual probability of non-performance. The annual probability of non-performance, Pfannual, is computed using the occurrence of earthquakes as a Poisson process [13, 22]:

$$Pf_{annual} = 1 - \exp\left(-\upsilon Pf_E\right) \rightarrow \beta_{annual} = -\Phi^{-1}\left(Pf_{annual}\right)$$

where βannual can be expressed as an reliability index as shown in Eq. (11), using the standard normal cumulative distribution function, Φ (·).

4. SEISMIC RESPONSES AND SSI SYSTEM

4.1. SEISMIC RESPONSES OF SSI SYSTEM

There are two main approaches for modeling and analyzing SSI systems, namely the direct method and the substructure method either in time domain or in frequency domain [6]. Considering the discretized dynamic equations of structure and soil simultaneously, the direct method models the soil and structure together, and the responses of soil and structure are determined simultaneously by analyzing SSI system in each time step [22].

In the direct method, the discretization of nonlinear dynamic equations can be expressed in FEM framework as:

$$\boldsymbol{M}\,\Delta\ddot{u} + \boldsymbol{C}\,\Delta\dot{u} + \boldsymbol{K}_{\mathrm{T}}\,\Delta\boldsymbol{u} = -\,\boldsymbol{m}_x\,\ddot{u}_{g,x}\left(t + \Delta t\right) - \boldsymbol{F}\left(t\right)$$

where M,C and KT are mass, damping, and tangent stiffness matrices of SSI model, respectively; Δu is the incremental vector of the relative displacements for SSI system between times tt and $t + \Delta t$; and F(t) is the vector of internal forces at time *t*. The term $\ddot{u}_{g,x}(t + \Delta t)$ is the free-field component of acceleration in x direction. The column matrix, $m_{x'}$ is the directional mass values of the structure only.

Over the past two decades, the use of damage and energy concepts for the seismic performance evaluation and design of structures has attracted considerable attention among the researchers [30, 44-47]. These concepts can be simultaneously used through a combined damage index namely Park-Ang damage index. The index is taken into account as a combined index, defined as the linear combination of the maximum displacement and the hysteretic energy dissipation for a structural element. For this reason, the damage index [44] is one of the indices that have widely been used for damage assessment and damage-based design of RC structures [14, 22, 28-30, 45-46, 48]. As shown in Table 1, some limit states of the performance levels depend on the damage indices.

An improved version of the index namely modified Park-Ang damage index [43] is defined based on the cross-section deformation of structural elements as:

$$DI = \frac{\theta_m - \theta_y}{\theta_u - \theta_y} + \frac{\beta}{\theta_u M_y} \int dE_H$$

where θmθm is the maximum rotation during loading history; θuθu and θyθy are the ultimate and yield rotation, respectively; My is the yield moment; and ∫dEH is the hysteretic energy dissipated in the same cross-section. Two connected indices, storey and overall damage indices, are computed using the weighting factors based on dissipated hysteretic energy at components and storey levels, respectively, as follows:

$$DI_{story} = \sum_{i=1}^{Ne} DI_i \cdot \lambda_{i,component} \quad ; \quad \lambda_{i,component} = \left[E_i / \sum E_i \right]_{component}$$

$$DI_{overall} = \sum_{i=1}^{Ns} DI_{i,story} \cdot \lambda_{i,story} \quad ; \quad \lambda_{i,story} = \left[E_i / \sum E_i \right]_{story}$$

where λi is energy weighting factor; and Ei is total absorbed energy by the component or storey *i*. *Ne*and *Ns* are the number of structural elements and stories, respectively.

4.2. FINITE ELEMENT MODEL OF SSI SYSTEM

OpenSEES [49], as an open-source computational software framework, is used for by simulation of SSI system, and performing nonlinear dynamic analyses of SSI system depicted in Fig. 1. Assuming materials of constant properties over its depth, soil encompasses different layers, and the foundation is considered as rigid strip footing. Beams and columns of structure are modeled using force-based nonlinear beam-column element with considering the spread plasticity along the element's length. The integration along each element is based on Gauss-Lobatto quadrature rule. Also, the infinite boundaries of soil are modeled using the artificial boundaries (Fig. 1). The model of soil-structure system shown inFig. 1 was successfully used by [14, 22, 31].

The Kent-Scott-Park model [50] is utilized for modeling the confined and unconfined concrete of cross-sections of structural elements. The constitutive parameters of this model are: *fc*=concrete peak strength in compression, *fu*=residual strength, ε0 =strain at peak strength, and εu =ultimate compressive strain (Fig. 2(a)). The material of cover and core concrete used in the cross-section are modeled as unconfined and confined, respectively. The constitutive model of confined concrete developed by Saatcioglu and Razvi [51], are used. Furthermore, to determine the ultimate compressive strain of confined concrete, the relationship introduced by Paulay and Priestley [52] is utilized as:

$$\varepsilon_{uc} = \varepsilon_{uo} + 1.4\frac{\rho_v f_{yh} \varepsilon_{us}}{f_{cc}}$$

where ε_{uc}, ε_{uo} and ε_{us} are the ultimate compressive strain of confined and unconfined concrete, and the ultimate strain of longitudinal steel reinforcement in tensile stress, respectively; ρ_v, f_{yh} and f_{cc} are the volumetric ratio, and the yield stress of confining steel reinforcement, and the peak strength of confined concrete in compression, respectively.

In this study, the one-dimensional J_2 plasticity model with linear hardening is utilized for modeling the constitutive behavior of the steel reinforcement. The material parameters defining J_2 plasticity model are: *fy*=yield strength, *H*=hardening modulus and *E*=Young's modulus (Fig. 2(b)).

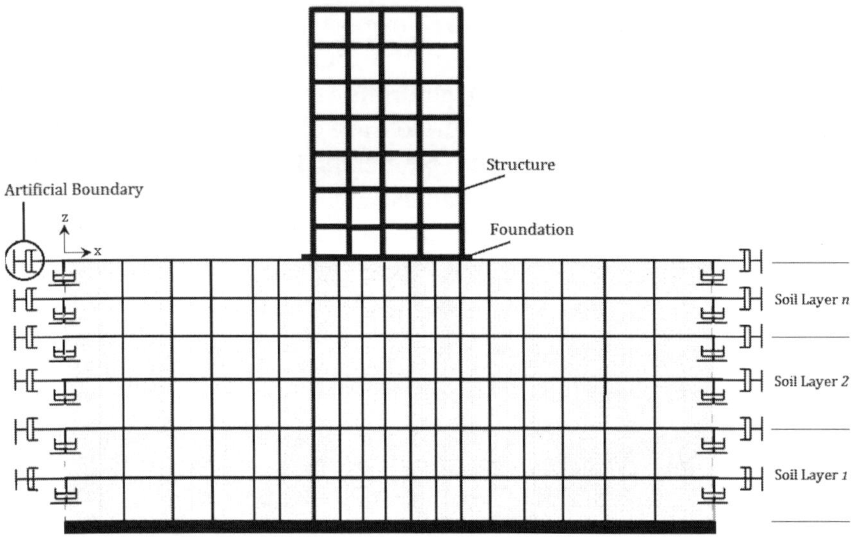

FIGURE 1. Direct method configuration for modelling of SSI system [22, 31]

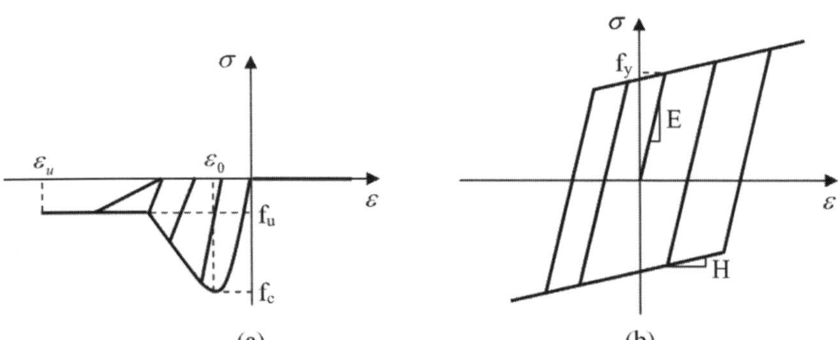

FIGURE 2. Material constitutive models; (a) Concrete, (b) Steel [22]

Soil layers are modeled using isoperimetric four-node quadrilateral finite elements and assuming bilinear displacement interpolation. The plane strain condition is assumed for the soil domain with considering a constant soil thickness corresponding to the inter-frame distance. The material of the soil is modeled using a modified pressure-independent multi-yield-surface J_2 plasticity model [53]. As shown in Fig. 3, this nonlinear model of soil material is described by a shear stress-strain backbone curve. The detailed description of the parameters of the shear stress-strain backbone curve can be found in [53].

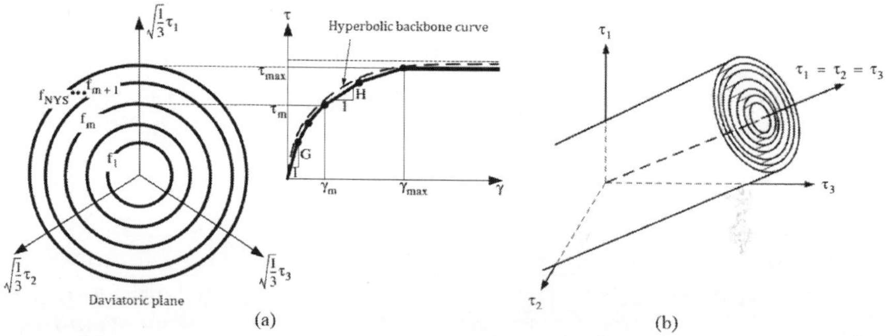

FIGURE 3. Yield surfaces of multi-yield-surface *J2* plasticity model; (a) Octahedral shear stress-strain, (b) Von Mises multi-yield surfaces [22, 53].

One of the major problems in SSI system for infinite media has been the modeling of the domain boundaries. Infinite boundaries have to absorb all outgoing waves and reflect no waves back into the computational domain. In this study, the standard viscous boundary proposed by Lysmer and Kuhlemeyer [54] is used for this purpose. This boundary can be described by two series of dashpots oriented normal and tangential to the boundary of a finite element mesh (Fig. 1) as follows:

$$C_n = a\rho V_p \quad ; \quad V_p = \sqrt{\frac{2G(1-v)}{\rho(1-2v)}}$$

$$C_s = b\rho V_s \quad ; \quad V_s = \sqrt{\frac{G}{\rho}}$$

where Cn and Cs are the normal and shear damping, respectively; ρ and v are the mass density and Poisson ratio of soil, respectively; *a* and *b* are dimensionless parameters to be determined, and Vp and Vs are dilatational and shear wave velocity of propagation, respectively. Standard viscous boundary with the normal and shear damping is modeled using the Zero Length element. Moreover, the parameters of soil layers are consisted of Gi as low strain shear modulus, Bi as bulk modulus, and τithat is shear strength, with *i*=1,2,...,*n* representing the soil layers' number. Other parameters of soil material depend on Bi,Gi, and Vp,iVp,i.

The material damping matrix, ***C***, of the SSI system is assembled by its corresponding damping matrices of structure and soil and considering the Rayleigh damping model. The factors of proportionality for damping matrices of structure and soil are computed based on 5% and 10% viscous damping respectively for structure and soil. The P-Δ effects are regarded in nonlinear time-history analyses. The accelerated Newton algorithm based on Krylov subspaces [55] is utilized for solving nonlinear equations of SSI system equilibrium.

5. ARTIFICIAL EARTHQUAKES

For RBDO of structures, it is then necessary to utilize accelerograms of compatible characteristics with a desired site. It is often difficult or impossible in some cases to choose a proper record for a site, since historically recorded accelerograms for a given site could be limited or scare. Hence, artificial earthquakes that are statistically influenced by desired properties of the given site are very useful for seismic design of structures. In this work, spectral representation method based on time domain procedure is used for the generation of synthetic ground motion records. The non-stationary ground motion is simulated using this method as [13]:

$$a(t)=I_m(t)\sum_{n=1}^{NFR}\left\{4S_{KT}(n\Delta f)\left[1+\delta_S R_N\right]\right\}\Delta f^{1/2}\sin(2\pi n\Delta f t+\theta_n)$$

where a(t),Im(t) and SKT(.) are the non-stationary ground motion, the modulation function and the specific power spectral density function (PSDF), respectively. NFR is the number of sine functions or frequencies included, between 0 and fmax, δS and RN are the coefficient of variation and a standard normal variable that used in ordinates of PSDF, Δf is frequency step, and θn are random phase angles with a uniform distribution between 0 and 2π2π. In this work, the modulation function expressed in [13] is used:

$$I_m(t)=\begin{cases}(t/T_1)^d & 0\le t\le T_1\\ 1 & T_1\le t\le T_2\\ e^{-c(t-T_2)} & T_2\le t\le T\end{cases}$$

where T1, T2 and T are specific times and the duration of the simulated record, d and c are constants. Also, the PSDF of the non-stationary ground motion suggested by Clough and Penzien [56] is considered as:

$$S_{KT}(f)=S_0\left[\frac{1+4\xi_g^2(f/f_g)^2}{\left(1-(f/f_g)^2\right)^2+4\xi_g^2(f/f_g)^2}\right]\left[\frac{(f/f_f)^4}{\left(1-(f/f_f)^2\right)^2+4\xi_f^2(f/f_f)^2}\right]$$

where S0 is the constant PSDF of input white-noise random process; fg and ξgξg are the characteristic ground frequency and the ground damping ratio; ff and ξfξare parameters for a high-pass filter to attenuate low frequency components. As listed in Table 2, the parameters for the generation of simulated ground motion are selected according to the values proposed by Möller et al. [13].

TABLE 2. Parameters for the generation of simulated ground motion (* aGaG = max(a(t))max(a(t)))

Parameter	$a_G \leq 350\, cm/s^2$ *	$350 \leq a_G \leq 700\, cm/s^2$	$a_G \geq 700\, cm/s^2$
T (sec)	5.12	10.24	20.48
T_1 (sec)	0.50	1.50	2.00
T_2 (sec)	4.00	8.00	16.00
c	2.0	1.0	0.7
d	2.0	2.0	2.0
NFR	100	200	300
f_{max} (Hz)	12	15	15
δ_S	0.40	0.40	0.40

The PGA values are obtained corresponding to hazard curves and produced for a specific region. As shown in Table 3, in this work the hazard curves presented by Möller et al. [13] are used. An artificial earthquake generated based on Eq. (19) is shown in Fig. 4.

TABLE 3. The PGA values of seismic hazard levels

Probability of exceedance	Recurrence interval	PGA (g)
50% in 50 years	73	0.27
10% in 50 years	475	0.6
5% in 50 years	975	0.8

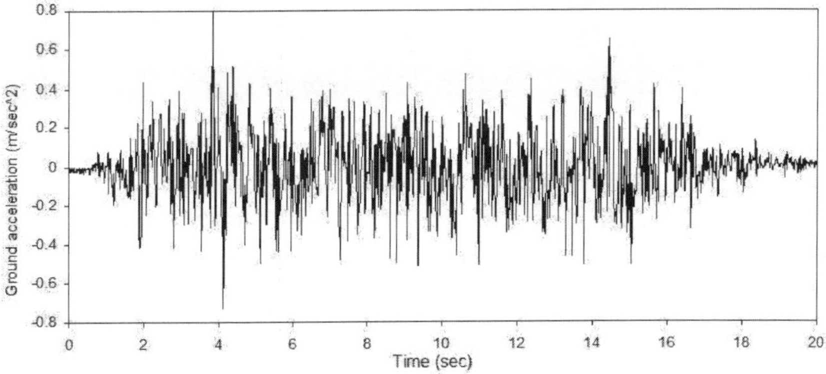

FIGURE 4. An artificial earthquake with PGA=0.8g [22]

6. THE NEW DISCRETE GRAVITATIONAL SEARCH ALGORITHM

Based on the work presented by Khatibinia *et al.* [14], a new discrete gravitational search algorithm (DGSA) based on the standard gravitational search algorithm (GSA) is utilized to find the optimum designs in the RBDO procedure.

6.1. GRAVITATIONAL SEARCH ALGORITHM

GSA was introduced by Rashedi *et al.* [23] as a new stochastic population based search algorithm based on the law of gravity and mass interactions. In GSA, each agent of the population represents a potential solution of the optimization problem. The *i*th agent in *t*th iteration is associated with a position vector, $X_i(t)=\{x_i^1, \ldots, x_i^d, \ldots, x_i^D\}$, and a velocity vector, $V_i(t)=\{v_i^1, \ldots, v_i^d, \ldots, v_i^D\}$. D is dimension of the solution space. Based on [23], the mass of each agent is calculated after computing the current population fitness for a minimization problem as follows:

$$M_i(t) = \frac{fit_i(t) - worst(t)}{\sum_{j=1}^{N}\left(fit_j(t) - worst(t)\right)}$$

where *N*, *Mi*(*t*) and fiti(t) represent the population size, the mass and the fitness value of agent *i* at *t*th iteration, respectively; and worst(t) is defined as the worst fitness of all agents.

To compute the acceleration of an agent, total forces from a set of heavier masses applied to it should be considered based on the law of gravity (Eq. (23)). Afterwards, the next velocity of an agent is calculated as a fraction of its current velocity added to its acceleration (Eq. (24)). Then, its next position could be calculated using Equation (25):

$$a_i^d(t) = \sum_{j \in kbest, j \neq i} rand_j G(t) \frac{M_j(t)}{R_{ij}(t) + \varepsilon}\left(x_j^d(t) - x_i^d(t)\right)$$

$$v_i^d(t+1) = rand_i v_i^d(t) + a_i^d(t)$$

$$x_i^d(t+1) = x_i^d(t) + v_i^d(t+1)$$

where adi, vdi and xdi present the acceleration, velocity and position of *i*th agent in dimension *d*, respectively. randi and randj are two uniformly distributed random numbers in the interval [0, 1]; ε is a small value; and Ri,j(t) is the Euclidean distance between agent *i* and *j*. *kbest* is the set of first *k*agents with the best fitness value and biggest mass, which is a function of time, initialized to k0 at the beginning and decreased with time. Here, k0 is set to *N* and is decreased linearly to 1.0. *G* is a decreasing function of time.

6.2. THE PROPOSED DISCRETE GSA

The binary GSA (BGSA) for solving discrete problem was developed by Rashedi *et al.* [25]. In the BGSA model, the positions of agents are indicated by one or zero. Hence, Eq. (24) can be used without any changes for updating the velocity of agents. Because of terms of probability, the velocity must be converted into interval [0, 1] by a transfer function, $S(v_i^d(t))$ [25]. For updating the position of the agents, Eq. (25) is re-defined as follows [25]:

$$x_i^d(t+1) = \begin{cases} complement\left(x_i^d(t)\right) & \text{for } r\ and < S\left(v_i^d(t)\right) \\ x_i^d(t) & \text{for } r\ and \geq S\left(v_i^d(t)\right) \end{cases}$$

Based on Eq. (26), a large computer memory is needed for the position of agents in BGSA. Also, coding and encoding of the position of agents is a time consuming process. In order to overcome the shortcomings of BGSA, a new DGSA based on the fundamental concept of the standard GSA with passive congregation is presented herein. The passive congregation strategy as perturbations operator can transfer information among agents in the optimization procedure [24]. Therefore, the search performance of DGSA can be improved using the passive congregation. To achieve this purpose, Khatibinia *et al.* [14] modified the velocity of agents based on the particle swarm optimizer with passive congregation (PSOPC) which is proposed by He *at al.* [24]. The modified velocity is expressed as follows:

$$v_i^d(t+1) = r_{i,1} v_i^d(t) + a_i^d(t) + r_{i,2}\left(p_{best,i}^d(t) - x_i^d(t)\right) + r_{i,3}\left(p_g^d(t) - x_i^d(t)\right) + r_{i,4}\left(p_i^d(t) - x_i^d(t)\right)$$

where pdbest, pdg and pdi are the best previous position of the *i*th agent, the best previous position among all the agents and a randomly selected agent from the population, respectively. $ri_{,1}$ and $ri_{,2}$ are the random number in the interval [0, 1], respectively; $ri_{,3}$ is the uniform random number in the interval (0, 1).

In DGSA, the scalar xdi∈{1,2,...,n} corresponds to discrete values of the set $\boldsymbol{A}$={*A1, A2,...,An*}. Hence, the position of agents is updated by the following equation instead of Eq. (26):

$$x_i^d(t+1) = INT\left(x_i^d(t) + v_i^d(t)\right)$$

Therefore, the coding and encoding of the position of agents are omitted; and the position of agents is calculated as the integer value. The current position of agents may violate from the values of the set $\boldsymbol{A}$. To avoid this problem, the current position of particles is limited as:

$$x_i^d(t+1) = \begin{cases} P_g^d & \text{if } fit\left(p_g^d(t)\right) = fit\left(p_g^d(t-1)\right) \\ P_{i,best}^d(t) & \text{if } fit\left(p_g^d(t)\right) \neq fit\left(p_g^d(t-1)\right) \\ INT\left(x_i^d(t) + v_i^d(t)\right) & \text{otherwise} \end{cases}$$

where INT(·) denotes the integral part function.

7. APPROXIMATING THE STRUCTURAL SEISMIC RESPONSES

MCS requires excessive computational cost for RBDO of structures in order to obtain an acceptable accuracy [11]. Because of the drawback, Khatibinia *et al.* [22] proposed a meta-model that predicted the mean and the standard deviation of the structural seismic responses in the RBDO process based on MCS. The meta-model consists of weighted least squares support vector machine (WLS-SVM) and wavelet kernel function, which is called WWLS-SVM [14, 22, 27].

7.1. WEIGHTED LEAST SQUARES SUPPORT VECTOR MACHINES

WLS-SVM was introduced as excellent machine learning algorithms in large-scale problems by Suykens *et al.* [26]. In fact, assigning weights to SVM as well as to the least squares version of SVM (LS-SVM) is resulted in more robust and precise prediction of functions [26].

$$\min J(w,e)=\frac{1}{2}\|w\|^2+\frac{1}{2}\gamma\sum_{i=1}^{n}\overline{v}_i e_i^2$$

WLS-SVM is described as the following optimization problem in primal weight space [26]:

Subject to the following equality constraints:

$$y_i = w^T\varphi(x_i)+b+e_i, \qquad i=1,2,\ldots,n$$

where $\{x_i, y_i\}_{i=1}^{n}$ is a training data set; $x_i \in R^n$ and $y_i \in R$ represent the input and output data. Operator $\varphi(.): R^n \rightarrow R^d$ is a function which maps the input space into a higher dimensional space. The vector, $w \in R^d$, represents weight vector in primal weight space. The symbols, $e_i \in R$ and $b \in R$, are the error variable and bias term, respectively. Using the optimization problem, Eq. (30), and the training data set, the WLS-SVM model could be expressed as:

$$y(x)=w^T\varphi(x)+b$$

It is impossible to indirectly compute ww from Eq. (30), for the structure of the function φ(x)φ(x) is generally unknown. Therefore, the dual problem shown in Eq. (30) is minimized by the Lagrange multiplier method as follows:

$$L(w,b,e;\boldsymbol{x}) = J(w,e) - \sum_{i=1}^{n} \alpha_i \left(w^T \varphi(\boldsymbol{x}_i) + b + e_i - y_i \right)$$

Based on the Karush-Kuhn-Tucker (KKT) conditions, by eliminating w and e the solution is given by the following set of linear equations:

$$\begin{bmatrix} \Omega + \boldsymbol{V}_\gamma & 1_n^T \\ 1_n & 0 \end{bmatrix} \begin{bmatrix} \alpha \\ b \end{bmatrix} = \begin{bmatrix} \boldsymbol{y} \\ 0 \end{bmatrix}$$

Where

$$V_\gamma = diag\{1/\gamma\bar{v}_1, \ldots, 1/\gamma\bar{v}_n\} \; ; \; \Omega_{i,j} = \langle \varphi(\boldsymbol{x}_i), \varphi(\boldsymbol{x}_j) \rangle_H \quad i, \; j = 1, \; \ldots, \; n$$

$$\boldsymbol{y} = [y_1, \; \ldots, \; y_n]^T; \; 1_n^T = [1, \; \ldots, \; 1] \; ; \; \boldsymbol{\alpha} = [\alpha_1, \; \ldots, \; \alpha_n]$$

According to Mercer's condition, a kernel K(.,.) is selected, such that:

$$K(\boldsymbol{x}, \bar{\boldsymbol{x}}) = \left\langle \varphi(\boldsymbol{x}), \varphi(\bar{\boldsymbol{x}}) \right\rangle_H$$

Consequently, the final WLS-SVM model for the prediction of functions becomes:

$$y(\boldsymbol{x}) = \sum_{i=1}^{n} \alpha_i K(\boldsymbol{x}_i, \boldsymbol{x}) + b$$

Weight $\bar{v}_k$ is estimated as follows [57-58]:

$$\overline{v}_k = \begin{cases} 1 & if\ |e_i / \hat{s}| \le c_1 \\ \dfrac{c_2 - |e_i / \hat{s}|}{c_2 - c_1} & if\ c_1 < |e_i / \hat{s}| \le c_2 \\ 10^{-4} & otherwise \end{cases}$$

where s^ is a robust estimation of the standard deviation for the error variables (ei=ai/D−1ii); constants c1 and c2 are typically chosen as c1=2.5 and c2=3. Here D−1ii denotes the *i*th primal diagonal element of inverse of matrix ***D***, which is the matrix on the left-hand of the system of the linearEq. (34) [59]. After that, weights v¯k are determined, and the model (Eq. (31)) is obtained following the solving WLS-SVM problem (Eq. (34)).

In WLS-SVM, Gaussian radial basis function (RBF) is frequently used as the kernel function, and it is expressed as:

$$K(\boldsymbol{x}, \overline{\boldsymbol{x}}) = \exp\left(-\frac{\|\boldsymbol{x} - \overline{\boldsymbol{x}}\|^2}{\sigma^2} \right)$$

where σ2 is a positive real constant usually called the kernel width.

Based upon the Suykens *et al.* [26], the WLS-SVM model based on RBF kernel function for predicting the output data is implemented using the following procedure:

Step 1. Assign training data $\{x_k, y_k\}_{k=1}^{Ntot}$, set N=$Ntot$.

Step 2. Find an optimum (γ,σ) combination on the all range of *Ntot* training data by 10-fold cross-validation, then solve linear system (Eq. (34)), and give the model (Eq. (37)).

Step 3. Sort the values |α|.

Step 4. Remove a small number of M points (typically 5% of the N points) which has the smallest values in the sorted $|\alpha|$.

Step 5. Retain N-M points and set $N=N-M$.

Step 6. Go to 2 and retrain on the reduced training set.

7.2. THE NEW META-MODEL-BASED WAVELET KERNEL

Wavelets as kernel function have been introduced and developed in ANNs and SVMs [60-63]. It has been shown that wavelet kernel functions are superior to other kernel functions in the training ANN and SVM. Accordingly, the kernel function of WLS-SVM is substituted with a specific kind of wavelet functions proposed by Khatibinia et al. [22]. The meta-model based on wavelet kernel function is called WWLS-SVM. The cosine-Gaussian Morlet wavelet is used as the kernel function of WLS-SVM. The wavelet function is mathematically written as follows:

$$\Psi(t)=\frac{1}{\sqrt{a}}\cos\left(\omega_0\left(\frac{t-b}{a}\right)\right)\exp\left(-0.5\left(\frac{t-b}{a}\right)^2\right)$$

where a and b are the scale factor and the translation factor, respectively.

According to Zhang *et al.* [64], the translation-invariant wavelet kernels can be explained as follows:

$$K(x,\bar{x})=\prod_{i=1}^{n}\Psi\left(\frac{x_i-\bar{x}_i}{a}\right)$$

where n is the number of samples; x and $\bar{x}\in R^n$

Therefore, according to Eqs. (40) and (41), the wavelet kernel function of the cosine-Gaussian Morlet wavelet is given as follows:

$$K(x,\bar{x})=\prod_{i=1}^{n}\frac{1}{\sqrt{a}}\cos\left(\omega_0\left(\frac{x_i-\bar{x}_i}{a}\right)\right)\exp\left(-0.5\frac{\|x_i-\bar{x}_i\|^2}{a^2}\right)$$

The accuracy of WWLS-SVM prediction depends on the good selection of its parameters. Selecting appropriate values of these parameters is important for obtaining the excellent predicting performance. Hence, in this study, GSA is used to find the WWLS-SVM optimal parameter, γ, and the wavelet kernel parameters, *a* and ω0. To achieve this purpose, a mean absolute percentage error (*MAPE*) is used to evaluate the performance of the WWLS-SVM model as follows:

$$MAPE=\frac{1}{n}\sum_{i=1}^{n}100\times\left|\frac{y_i-\bar{y}_i}{y_i}\right|$$

where *y* and $\bar{y}$ are the actual value and the predicted value, respectively.

The WWLS-SVM training stage during GSA is performed according to the *k*-fold cross-validation (CV) [26]. Consequently, an optimization problem based on *MAPE* of the *k*-fold CV (*MAPEk-fold CV*) is expressed as:

$$\begin{aligned}&\text{Minimize} \quad MAPE_{k-fold\,CV}\\&\text{Subject to} \quad \gamma_{\min}\le\gamma\le\gamma_{\max}\\&\qquad a_{\min}\le a\le a_{\max}\\&\qquad \omega_{0\min}\le\omega_0\le\omega_{0\max}\end{aligned}$$

The converged solution is affected by the setting value of parameters in GSA. In this study, the values are selected based on the general recommendations by Rashedi *et al.* (2009). The flowchart of the meta-model based on WWLS-SVM [22] and GSA [14, 23] is shown in Fig. 5.

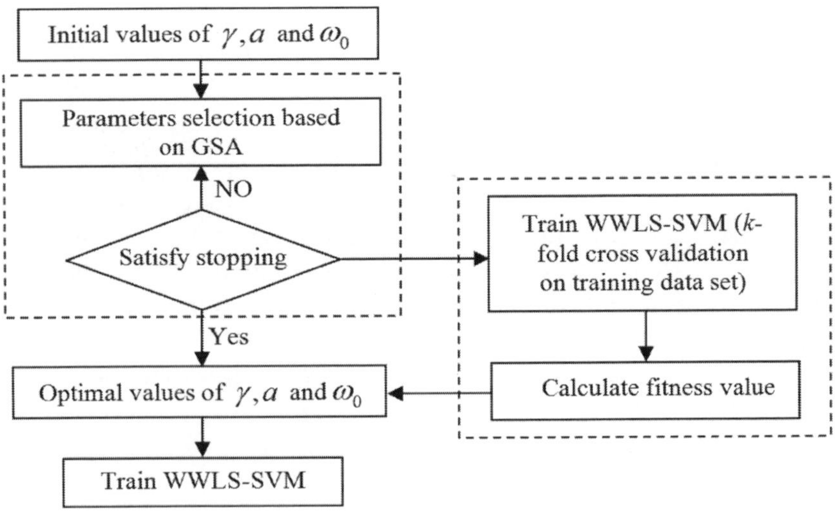

FIGURE 5. The prediction meta-model based on WWLS-SVM and GSA [14]

8. PREDICTING FAILURE PROBABILITY OF STRUCTURES

In the RBDO procedure, nonlinear time-history analysis of SSI system is used and it may be failed regarding a number of random structures [65]. In fact, a number of structures collapse and then lose their stability. Hence, these structures should be identified and eliminated from optimization process. For this purpose, a failure probability is considered as stability criterion. An efficient method is presented to train the failure probability with high performance [65]. This efficient method is consisted of a modified adaptive neuro fuzzy inference system (ANFIS) with a hybrid of fuzzy c-means (FCM) [66] and fuzzy particle swarm optimization (FPSO) [67]. To train the modified ANFIS, the input–output data are classified by a hybrid algorithm consisting of FCM-FPSO clustering. The optimum number of ANFIS fuzzy rules is determined by subtractive algorithm (SA).

8.1. HYBRID OF FCM AND FPSO FOR CLUSTERING

The FCM algorithm has been extensively studied and is known to converge to a local optimum in nonlinear problems. Moreover, the FPSO algorithm is robust method to increase the probability of achieving the global optimum in comparison with the FCM algorithm. The FCM algorithm is faster than the FPSO algorithm because it requires fewer function evaluations. This shortcoming of FPSO can be dealt with selecting an adequate initial swarm [65].

In this study, a hybrid clustering algorithm called FCM-FPSO is presented to use the merits of both FCM and FPSO algorithms and increase the procedure of convergence. In this way, the FCM algorithm finds an adequate initial swarm FPSO algorithm for commencing the FPSO. For this purpose, first, the FCM algorithm is utilized to find a preliminary optimization that shown by XFCM. This optimum solution is copied *NFCM* times to create the some part of the initial swarm FPSO. Other particles of the initial swarm, i.e. $X_{rnd,\,j}\,(j=1,\,2,\,\ldots,\,N_{PSO}-N_{FCM})$, are selected randomly to complete the initial swarm. Then, the FPSO algorithm is used using this initial swarm. The algorithm flow of the FCM–FPSO strategy is shown in Fig. 6.

8.2. MODIFIED ANFIS

An ANFIS model depends on the number of ANFIS fuzzy rules and membership functions. In other words, creating an ANFIS model with a minimum number of fuzzy rules can eliminate a well-known drawback. Therefore, for overcoming of this drawback, Khatibinia *et al.* [65] proposed a modified ANFIS to predict the probability of failure. In this model, the number of clusters, the cluster centers and membership grades are considered as parameters which optimized by subtractive algorithm (SA) and the hybrid of FCM-FPSO and used in FIS for tuning ANFIS. The algorithm flow of the proposed model is shown in Fig. 7. The proposed method is executed in the following steps [65]:

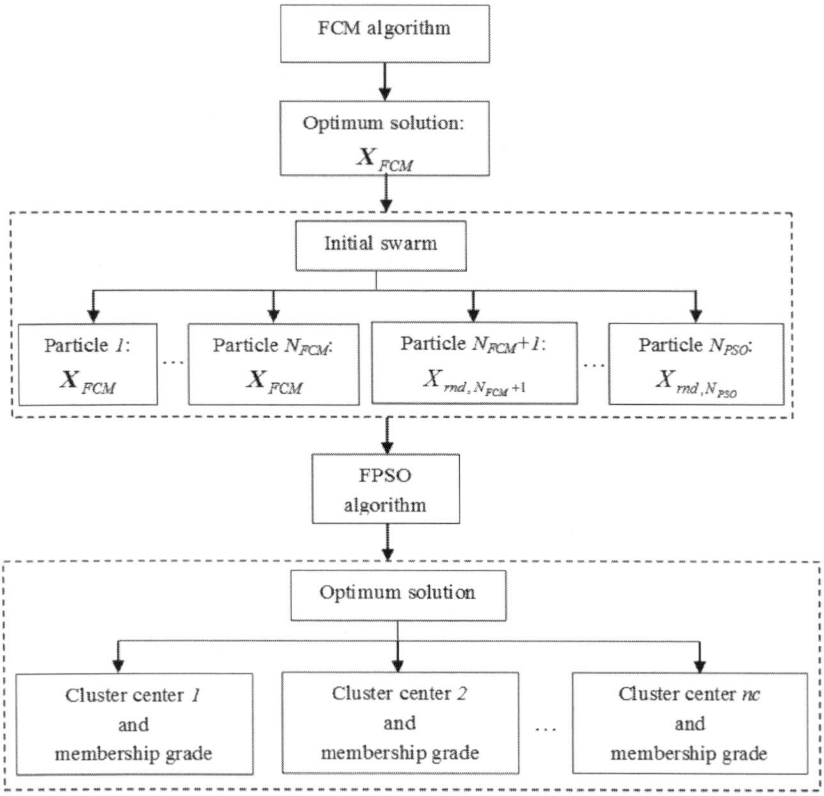

FIGURE 6. The algorithm flow of the FCM–FPSO for clustering [65]

Step 1. SA finds the optimum number of the clusters (*nc*).

Step 2. The hybrid FCM-FPSO algorithm partitions training data to *nc* clusters and determines membership grades each of clusters. This parameters is used for optimizing the center of rules and membership functions for the input and output data.

Step 3. The FIS structure with a minimum number of fuzzy rules and membership functions is generated by using SA and the hybrid FCM-FPSO algorithm. The FIS uses Gaussian function and linear function for membership function of input and output, respectively. These parameters are tuned for the ANFIS.

Step 4. The ANFIS is employed for training data. The ANFIS uses a hybrid learning algorithm to identify parameters of Sugeno-type fuzzy inference systems. a combination of the least-squares method and the back-propagation gradient descent method are applied for training FIS membership functions.

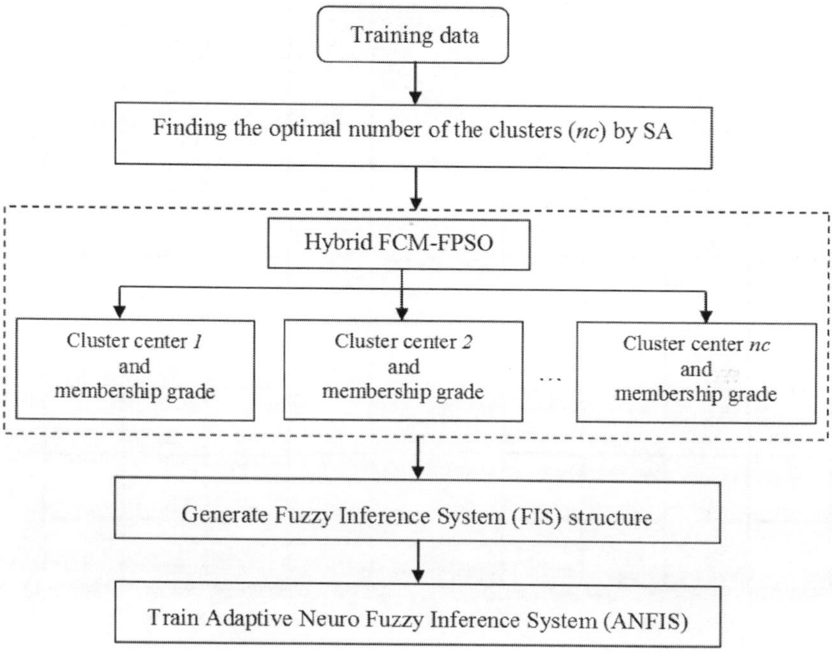

FIGURE 7. The algorithm flow of the proposed method [65]

9. NUMERICAL EXAMPLES

In this work, two RC frame structures shown in Fig. 8 are selected as illustrative examples. Three layers of sand associated with material properties varying over its depth are considered as the soil under the frames. The depth of each soil layer and the entire width of soil domain are considered to be 10 m and 100 m, respectively. The soil is also assumed to have plane strain condition with a constant thickness of 5.0 m in proportion to the inter-frame distance. Inertia properties of the soil mesh are considered using lumped mass matrices modeling with soil mass density of 17 kN/m³ for all soil layers. The values of the dead and live loads are considered to be DL=5.884 N/mm² (600 kg/m²) and LL=1.961 N/mm² (200 kg/m²), respectively.

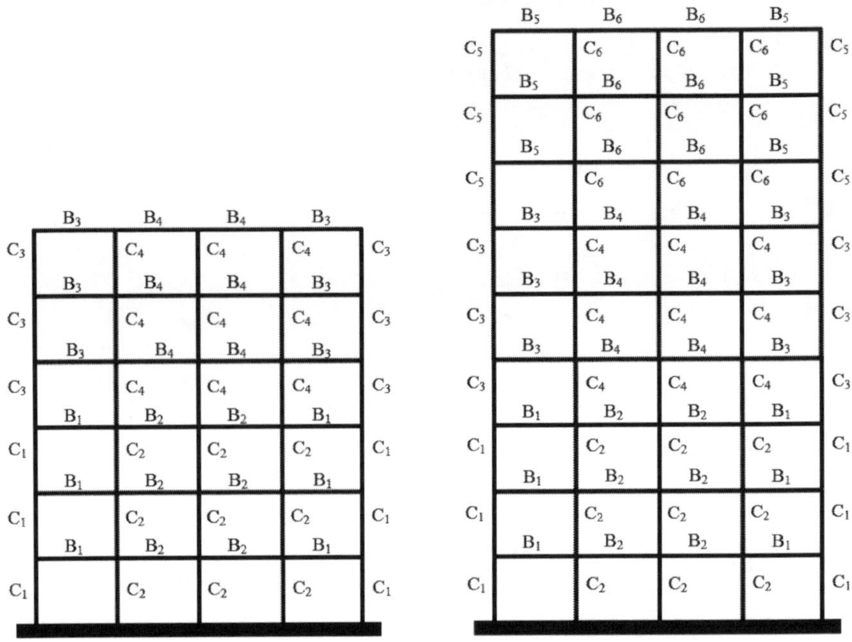

FIGURE 8. Two illustrated RC frame structures [22]

For vertical continuity on the dimensions along the height of a column, the section database of columns is divided into three types in the height of RC frame. Hence, a database shown in Table 4 is generated. Similarly, the section database of beams is divided into three types in the height of RC frame. Distribution of beam dimensions along the height of the frame is shown in Table 4. The diameter of longitudinal bars for beams and columns is laid between 12 mm and 32 mm in the databases.

The initial cost is calculated for wC=60w and wS=700w Euros. To calculate the total expected cost of repair, first, the cumulative distribution is obtained using MCS and the proposed meta-model for the response *DIoverall* adjusting a Beta distribution. Then, the density function is assigned by the derivative of the cumulative distribution. The target values of reliability indices corresponding to the three performance levels (Table 1) considered in RBDO of RC structures are equal to 1.276, 2.326 and 2.697, respectively [52]. For RBDO of RC structures, the probability density function (PDF), the mean and standard deviation (SD) values for each random constitutive parameter are listed in Table 5.

The concrete material parameters shown in Table 5 are considered for the cover of column cross-sections. The strain corresponding to the peak strength, $\varepsilon 0o$, and the residual strength, fu_o, for unconfined concrete are selected as 0.002 and 0.0, respectively. Shear-wave velocity, *Vs,* and friction angle, φ, are considered as random parameters of the soil layers. The other parameters of soil layers depend on their shear-wave velocities. Thus, in the process, first, the shear-wave velocities of soil layers are randomly selected; then, the other parameters are computed based on the shear-wave velocity. The PGA value, *ag*, and the central frequency, *fg*, for soil filter are considered as random variables in the RBDO process. The PGA value is also taken into account as follows [13, 22]:

$$a_g = \bar{a}_g\left(1 + \sigma_{\bar{a}g}\right)$$

where a¯g and σa¯g are the mean and the standard deviation of PGA, respectively. The values of these parameters are shown in Table 6.

The presented DGSA requires the user to specify several internal parameters that can affect convergence behavior at the search space. It is found that a population of 50 agents can be adequate. Higher values are not recommended, as this will increase significantly computation time in RBDO. In addition, different optimization runs are carried out for RBDO model in this study, so optimum designs are found by DGSA about 150 iterations. Due to the effect of decreasing gravity, the actual value of the gravitational

TABLE 4. The section database of columns and beams. (a. The unit of sections is cm)

Number	Column (h × h)[a]			Beam (h × b)[a]		
	Type 1	Type 2	Type 3	Type 1	Type 2	Type 3
1	65 × 65	55 × 55	45 × 45	55 × 45	50 × 45	45 × 40
2	60 × 60	50 × 50	40 × 40	55 × 40	50 × 40	45 × 35
3	55 × 55	50 × 50	40 × 40	55 × 35	50 × 35	45 × 30
4	55 × 55	45 × 45	40 × 40	55 × 30	50 × 30	45 × 25
5	55 × 55	45 × 45	35 × 35	50 × 40	45 × 40	40 × 35
6	50 × 50	45 × 45	40 × 40	50 × 35	45 × 35	40 × 30
7	55 × 55	45 × 45	35 × 35	50 × 30	45 × 30	40 × 30
8	-	-	-	50 × 30	45 × 30	35 × 30

constant, $G(t)$, depends on the actual age of the universe. In this study, $G(t)$ is considered as a linear decreasing function [23] in DGSA. Since the initial value of the gravitational constant is found to affect the optimization results significantly, the fixed value of $G0 = 50$ is utilized in this study. In order to consider the stochastic nature of the optimization process, ten independent optimization runs are performed and the best solution is considered as the final results.

9.1. EXAMPLE 1: SIX-STOREY RC FRAME

Six-storey RC frame is shown in Fig. 8(a). In the frame, the length of each bay and the height of stories are 5 m and 3 m, respectively. The members of the structure are divided into four groups for the columns C_1, C_2, C_3, C_4 and four groups B_1, B_2, B_3 and B_4 for the beams. The groups of structural elements are presented in Fig. 8(a).

TABLE 5. The marginal probability distribution, mean and standard deviation of materials

Material	Variable	PDF	Mean	SD
Concrete				
	f_{co} (Mpa)	Lognormal	28	2
	ε_{uo}	Normal	0.0035	0.00035
Steel				
	f_y (Mpa)	Lognormal	340	25
	E (Mpa)	Lognormal	210000	8000
	H	Normal	0.015	0.0015
Soil				
	V_{s1} (m/s)	Normal	375	10
	φ_1	Normal	37.5	1.0
	V_{s2} (m/s)	Normal	300	20
	φ_2	Normal	37.5	1.0
	V_{s3} (m/s)	Normal	200	10
	φ_3	Normal	32.5	1.0

TABLE 6. Properties of the random variables for generation of the artificial earthquakes

Variable	PDF	Mean	SD
$\bar{a}_g$ (cm/s^2)	Lognormal	300	100
$\sigma_{\bar{a}g}$	Normal	-	0.15
f_g (Hz)	Normal	2.5	0.375

9.1.1. TRAINING AND TESTING THE META-MODEL

To predict the mean, $\bar{R}_i$, and the standard deviation, σRi, of *i*th seismic response during RBDO of the frame, the proposed meta-models are trained based on the generated database. The WWLS-SVM training during GSA is performed according to five-fold cross-validation. The lower and upper bonds of the parameters required in the optimization process are selected as $\gamma \in [1.0,\ 500]$, $a \in [0.5,\ 5.0]$ and $\omega_0 \in [1.0,\ 10]$, resp ectively. Therefore, the training optimal parameters of the meta-model associated with the mean and the standard deviation of seismic responses are shown in Table 7.

In order to validate the performance and accuracy of the proposed meta-model, relative root-mean-squared error, i.e. *RRMSE*, and R^2 as the absolute fraction of variance, during testing the meta-model and WLS-SVM, are defined using the following equations:

$$RRMSE = \sqrt{\frac{n\sum_{i=1}^{n}\left(y_i - \bar{y}_i\right)^2}{\left(n-1\right)\sum_{i=1}^{n} y_i^2}}$$

$$R^2 = 1 - \left(\frac{\sum_{i=1}^{n}(y_i - \bar{y}_i)^2}{\sum_{i=1}^{n}\bar{y}_i^2}\right)$$

TABLE 7. Optimal parameters of the meta-model for training the mean and the standard deviation of the seismic responses

Parameter	Mean			Standard deviation		
	γ	a	ω_0	γ	a	ω_0
u_{max}	375.73	2.892	4.274	284.62	1.647	4.048
DR_{max}	386.04	3.804	6.347	314.42	2.104	4.702
DIL_{max}	400.48	3.615	5.895	373.67	2.548	3.692
$DI_{overall}$	365.37	4.052	6.329	308.38	1.947	4.082

The smaller *RRMSE* and *MAPE* and the larger R^2 are indicative of better performance generality. The comparison of WWLS-SVM and WLS-SVM, with respect to *MAPE*, *RRMSE*, and R^2 is shown in Table 8.

TABLE 8. Performance associated with the mean and the standard deviation of the seismic responses.

Method	Statistical parameters	Mean				Standard deviation			
		u_{max}	DR_{max}	DIL_{max}	$DI_{overall}$	u_{max}	DR_{max}	DIL_{max}	$DI_{overall}$
WWLS-SVM									
	MAPE	2.008	2.417	2.337	2.172	2.163	2.075	2.532	3.406
	RRMSE	0.018	0.022	0.020	0.021	0.023	0.024	0.045	0.028
	R^2	0.9999	0.9998	0.9998	0.9999	0.9999	0.9998	0.9998	0.9997
WLS-SVM									
	MAPE	5.328	6.302	5.392	5.862	5.386	6.017	6.007	5.737
	RRMSE	0.057	0.074	0.059	0.061	0.079	0.094	0.091	0.064
	R^2	0.9988	0.9979	0.9987	0.9987	0.9985	0.9984	0.9988	0.9988

As given in Table 8, the proposed meta-model trained for the mean and the standard deviation of seismic responses has proper performance generality. Thus, the approximating performance of the meta-model based on WWLS-SVM and GSA is better than the WLS-SVM with RBF kernel in predictive ability and precision.

9.1.2. RESULTS OF RBDO

In this example, RBDO of the RC frame is performed using DGSA associated with WWLS-SVM-based MCS. In the reliability process, the reliability indices, βannual, are estimated using WWLS-SVM-based MCS with 10^6 samples generated with the LHD method. The cross-section of beams and columns are selected from Types 2 and 3, which are shown in Table 4. The optimum designs of the RC frame are listed in Table 9. Furthermore, the optimal solutions of DGSA are also compared with those of BGSA in Table 9.

As shown in Table 9, the optimal solutions of DGSA are better than those of BGSA in terms of the total cost and the number of iterations. The minimum reliability index, βannual, obtained corresponding to each performance level by DGSA and BGSA is shown in Table 9.

The convergence histories of the optimum objective function are shown in Fig. 9 for DGSA and BGSA models. As can be seen in Fig. 9, DGSA method is more efficient than BGSA method. Optimum designs are found by DGSA and BGSA in 4450 and 5900 required approximate analyses by the meta-model, respectively.

TABLE 9. Optimum designs obtained by DGSA and BGSA

Element groups no.	DGSA		BGSA	
	Cross-section	ρ (%)	Cross-section	ρ (%)
C_1	55 × 55	2.67	55 × 55	3.23
C_2	55 × 55	2.64	55 × 55	3.20
C_3	45 × 45	2.56	45 × 45	2.60
C_4	45 × 45	2.33	45 × 45	2.48
B_1	50 × 45	1.98	50 × 45	2.38
B_2	50 × 40	2.13	50 × 40	2.30
B_3	45 × 40	1.81	45 × 40	1.88
B_4	45 × 35	1.69	45 × 35	1.78
C_{IC} (Euro)	3448		3542	
C_{RC} (Euro)	1080		1045	
C_{TOT}(Euro)	4528		4587	
Iterations	89		118	
$\beta_{annual}^{Operational}$	1.4435		1.6901	
$\beta_{annual}^{Life\ safety}$	2.4885		2.7342	
$\beta_{annual}^{Collapse}$	2.7986		3.1384	

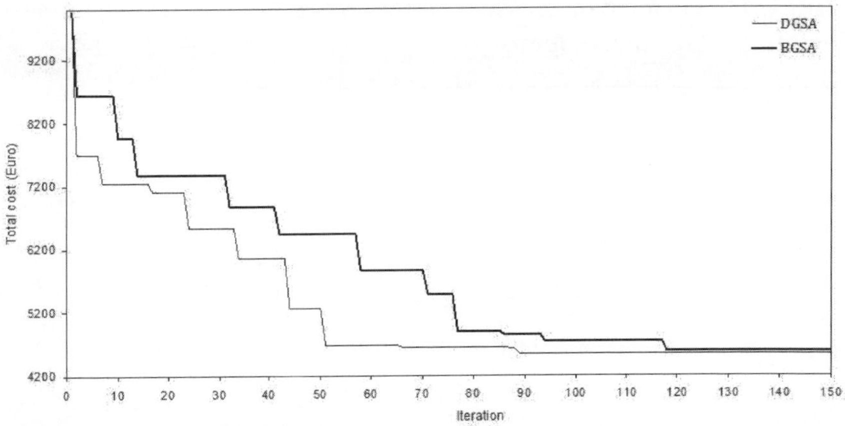

FIGURE 9. Convergence histories of the best solution of DGSA and BGSA

9.2. EXAMPLE 2: NINE-STOREY RC FRAME

Nine-storey RC frame is shown in Fig. 8(b). In the frame, the length of each bay and the height of stories are 5 m and 3 m, respectively. The members of the structure are divided into six groups for the columns and six groups for the beams. The groups of structural elements are presented in Fig. 8(b).

9.2.1. TRAINING AND TESTING THE META-MODEL

After training database using the presented WWLS-SVM optimal parameters of the meta-model associated with the mean and the standard deviation of seismic responses are shown in Table 10. Furthermore, the performance generality of the proposed meta-model and WLS-SVM is given in Table 10 in terms of *MAPE, RRMSE* and R^2.

TABLE 10. Performance associated with the mean and the standard deviation of seismic responses.

Method	Statistical parameters	Mean				Standard deviation			
		u_{max}	DR_{max}	DIL_{max}	$DI_{overall}$	u_{max}	DR_{max}	DIL_{max}	$DI_{overall}$
WWLS-SVM									
	MAPE	2.837	3.028	3.127	2.689	3.024	2.682	3.008	3.105
	RRMSE	0.026	0.038	0.035	0.023	0.032	0.024	0.031	0.036
	R^2	0.9999	0.9998	0.9998	0.9999	0.9998	0.9999	0.9998	0.9998
WLS-SVM									
	MAPE	5.538	6.346	6.483	6.006	6.305	5.396	5.843	5.579
	RRMSE	0.094	0.138	0.162	0.105	0.1057	0.0987	0.0998	0.0924
	R^2	0.9987	0.9985	0.9981	0.9988	0.9987	0.9988	0.9988	0.9988

The results of Table 10 demonstrate that the meta-model is better than the WLS-SVM method in terms of performance generality. Therefore, the meta-model is reliably employed to predict the necessary responses during the RBDO process.

9.2.2. RESULTS OF RBDO

As the first example, in this example RBDO of the RC frame is performed using DGSA and BGSA associated with WWLS-SVM-based MCS. In this example, the cross-section of beams and columns are selected from Types 1, 2 and 3, which are shown in Table 4. The best optimum designs of the RC frame are listed in Table 11.

As revealed in Table 11, the optimal solutions of DGSA are better than those of BGSA in terms of the total cost and the number of iterations. The minimum reliability index, βannual, obtained corresponding to each performance level by DGSA and BGSA is shown in Table 11.

The convergence histories of the optimum objective function are shown in Fig. 10 for DGSA and BGSA models. As can be seen in Fig. 10, DGSA method is more efficient than BGSA method. Optimum designs are found by DGSA and BGSA in 4150 and 6150 required approximate analyses by the meta-model, respectively.

TABLE 11. Optimum designs obtained by DGSA and BGSA

Element groups No.	DGSA		BGSA	
	Cross-section	ρ (%)	Cross-section	ρ (%)
C_1	65 × 65	2.49	65 × 65	2.90
C_2	60 × 60	2.50	60 × 60	3.08
C_3	55 × 55	2.53	55 × 55	2.87
C_4	50 × 50	2.49	50 × 50	2.79
C_5	45 × 45	2.27	45 × 45	2.68
C_6	40 × 40	2.29	40 × 40	2.89
B_1	55 × 45	1.88	55 × 45	2.18
B_2	55 × 40	1.80	55 × 40	1.98
B_3	50 × 45	1.82	50 × 45	2.01
B_4	50 × 40	1.87	50 × 40	1.98
B_5	45 × 40	1.68	45 × 40	1.79
B_6	45 × 35	1.66	45 × 35	1.85
C_{IC} (Euro)	5571		5736	
C_{RC} (Euro)	1004		987	
C_{TOT}(Euro)	6575		6723	
Iterations	83		123	
$\beta_{annual}^{Operational}$	1.4985		1.7101	
$\beta_{annual}^{Life\ safety}$	2.4953		2.7996	
$\beta_{annual}^{Collapse}$	2.8514		3.1837	

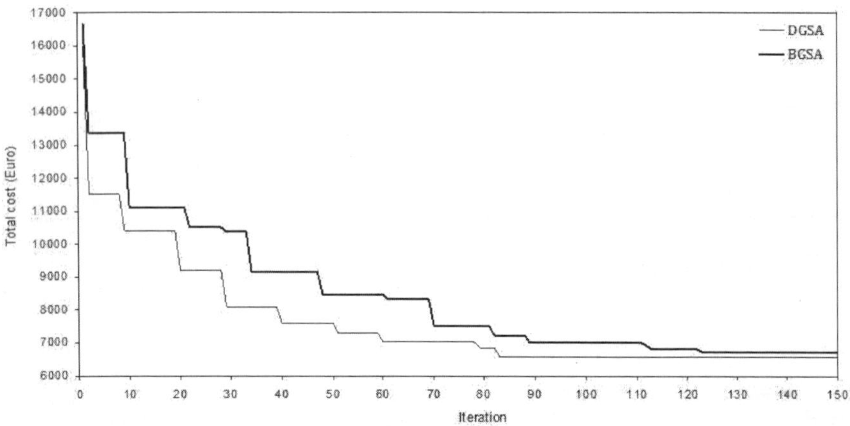

FIGURE 10. Convergence histories of the best solution of DGSA and BGSA

10. CONCLUSIONS

In general, the optimum design of structures depends on a number of parameters that are inherently uncertain. Reliability-based design optimization (RBDO) has been employed as the only method that assesses the influence of uncertain parameters and balance both cost and safety of structures. To account for all necessary uncertain and random parameters in RBDO of RC structures and to achieve the realistic optimum design of RC structures, the uncertain material properties of soil and structure, as well as the characteristics of ground motions should be considered as random parameters. Furthermore, the realistic seismic responses of RC structures can be account by consideration of soil-structure interaction (SSI) effects. In this work, a new discrete gravitational search algorithm (DGSA) and a new meta-modeling framework were incorporated for RBDO of RC structures with Performance-Based Design (PBD) under seismic loading. The objective of the RBDO problem was to minimize the total cost whereas the deterministic constraints and the system reliability index corresponding to each of the performance levels should not exceed a specified target. Based on this study, the following conclusions can be drawn:

- To reduce the computational effort and computational cost of the Monte-Carlo Simulation (MCS) method, a new meta-model based on a wavelet weighted least squares support vector machine (WWLS-SVM) and gravitational search algorithm (GSA) was utilized in the RBDO procedure. Therefore, the proposed meta-model, as a substitute for the nonlinear dynamic analysis of SSI system, can estimate the reliability index through MCS with a small computational cost.

- The WWLS-SVM and kernel parameters were simultaneously optimized in the proposed meta-model in order to improve performance generality of WWLS-SVM. Numerical results of training and testing the meta-model indicated that performance generality of the meta-model was higher in comparison to WLS-SVM. Hence, the proposed meta-model can predict the nonlinear dynamic analysis of SSI system in terms of accuracy and flexibility.

- The proposed DGSA was presented based on the standard GSA with passive congregation. The passive congregation strategy can be considered as perturbations operator in the optimization procedure. Therefore, the presented DGSA using the passive congregation can transfer information among agents avoiding local minima. Furthermore, the coding and encoding of the position of agents as a time consuming process is omitted in DGSA. To eliminate this drawback, the position of agents was calculated as the integer value. The optimum designs obtained by DGSA were compared with those produced by BGSA model. Numerical examples showed that the proposed DGSA can converge and reach the optimum design more quickly than the BGSA model.

Future extension of current research could include reducing the computations involved in the PBD by replacing MCS with the response surface method or the importance sampling technique. The constraints imposed on the objective function could be also treated as random quantities (see [68]).

11. ACKNOWLEDGEMENTS

Our special thanks go to Dr. Eysa Salajegheh (Distinguished Professor of Structural Engineering) in Department of Civil Engineering at Shahid Bahonar University of Kerman, Iran, for his cooperation in this research work.

REFERENCES

1. Moustafa, A. (Ed.). (2013). *Earthquake-resistant structures: design, assessment and rehabilitation*, InTech, Croatia.

2. Takewaki, I.; Fujita, K. & Moustafa, A. (2013). *Improving the earthquake resilience of buildings: the worst case approach*, Pham, H. (Ed.), Springer Series in Reliability Engineering, London.

3. SEAOC, Vision 2000 Committee, (1995). *Performance based seismic engineering of buildings*. Structural Engineers Association of California, Sacramento, CA.

4. ATC40, (1997). *Seismic evaluation and retrofit of concrete buildings*, Applied Technology Council, California Seismic Safety Commission.

5. FEMA-356, (2000). *Pre standard and commentary for the seismic rehabilitation of buildings*. Federal Emergency Manage Agency, prepared by the American Society of Civil Engineers, Washington.

6. Wolf, J.P. & Song, Ch. (2002). Some cornerstones of dynamic soil-structure interaction, *Engineering Structures*, Vol. 24, pp. 13-28.

7. Schuëller, G.I. & Jensen, H.A. (2008). Computational methods in optimization considering uncertainties-an overview. *Computer Methods in Applied Mechanics and Engineering*, Vol. 198(1), pp. 2–13.

8. Park, G.J.; Lee, T.H.; Lee, K.H. & Hwang, K.H. (2006). Robust design: an overview, *AIAA Journal*, Vol. 44(1), pp. 181-191.

9. Lopez, R.H.; Lemosse, D.; Cursi, J.E.S.; Rojas, J. & El-Hami, A. (2011). An approach for the reliability based design optimization of laminated composites, *Engineering Optimization*, Vol. 43(10), pp. 1079-1094.

10. Foley, C.M.; Pezeshk, S. & Alimoradi, A. (2007). Probabilistic performance-based optimal design of steel moment-resisting frames, I: formulation,*Journal of Structural Engineering*, ASEC, Vol. 133(6), pp. 757-766.

11. Lagaros, N.; Garavelas, A. & Papadrakakis, M. (2008). Innovative seismic design optimization with reliability constraints, *Computer Methods in Applied Mechanics and Engineering*, Vol. 198, pp. 28-41.

12. Fragiadakis, M. & Papadrakakis, M. (2008). Performance-based optimum seismic design of reinforced concrete structures, *Earthquake Engineering and Structural Dynamics*, Vol. 37, pp. 825-844

13. Möller, O.; Foschi, R.; Rubinstein, M. & Quiroz, L. (2009). Structural optimization for performance-based design in earthquake engineering: applications of neural networks, *Structural Safety*, Vol. 31, pp. 490-499.

14. Khatibinia, K.; Salajegheh, J.; Fadaee, M.J. & Salajegheh, E. (2013). Seismic reliability assessment of RC structures including soil-structure interaction using wavelet weighted least squares support vector machine, *Reliability Engineering and System Safety*, Vol. 110, pp. 22-33.

15. Salajegheh, E.; Gholizadeh, S. & Khatibinia, M. (2008). Optimal design of structures for earthquake loads by a hybrid RBF-BPSO method,*Earthquake Engineering and Vibration Engineering*, Vol. 7, pp. 14-24.

16. Gholizadeh, S. & Salajegheh, E. (2009). Optimal design of structures subjected to time history loading by swarm intelligence and an advanced meta-model, *Computer Methods in Applied Mechanics and Engineering*, Vol. 198, pp. 2936-2949.

17. Chen, K.Y. (2007). Forecasting systems reliability based on support vector regression with genetic algorithms, *Reliability Engineering and System Safety*, Vol. 92, pp. 423-32.

18. Zhiwei, G. & Guangchen B. (2009). Application of least squares support vector machine for regression to reliability analysis, *Chinese Journal of Aeronautics*, Vol. 22, pp. 160-166.

19. Tan, X.H.; Bi, W.H.; Hou, X.L. & Wang, W. (2011). Reliability analysis using radial basis function networks and support vector machines,*Computers and Geotechnics*, Vol. 38, pp. 178-86.

20. Moura, M.C.; Zio, E.; Lins, I.D. & Droguett, E. (2011). Failure and reliability prediction by support vector machine regression of time series data,*Reliability Engineering and System Safety*, Vol. 96, pp. 1527-34.

21. Dai, H.; Zhang, H. & Wang, W. (2012). A support vector density-based importance sampling for reliability assessment, *Reliability Engineering and System Safety*, Vol. 106, pp. 86-93.

22. Khatibinia, M.; Salajegheh, E.; Salajegheh, J. & Fadaee M.J. (2013). Reliability-based design optimization of RC structures including soil-structure interaction using a discrete gravitational search algorithm and a proposed meta-model, *Engineering Optimization*, Vol. 45 (10), pp. 1147-1165.

23. Rashedi, E.; Nezamabadi-pour, H. & Saryazdi, S. (2009). GSA: a gravitational search algorithm, *Information Science*, Vol. 179(13), pp. 2232-2248.

24. He, S.; Wu, Q.H.; Wen, J.Y.; Saunders, J.R. & Paton, R.C. (2004). A particle swarm optimizer with passive congregation, *BioSystems*, Vol. 78, pp. 135-147.

25. Rashedi, E.; Nezamabadi-pour, H. & Saryazdi, S. (2010). BGSA: binary gravitational search algorithm, *Natural Computing*, Vol. 9(3), pp. 727-745.

26. Suykens, J.A.K.; Brabanter, J.D.; Lukas, L. & Vandewalle, J. (2002). Weighted least squares support vector machines: robustness and sparse approximation, *Neurocomputing*, Vol. 48, pp. 85-105.

27. Gharehbaghi, S. & Khatibinia, M. (2015). Optimal seismic design of reinforced concrete structures subjected to time-history earthquake loads using an intelligent hybrid algorithm. *Earthquake Engineering and Engineering Vibration*, Vol. 14(1), pp. 97-109.

28. Gharehbaghi, S. & Salajegheh, E. (2011). Optimum distribution of hysteretic energy in reinforced concrete structures, *proceeding of the 6th national conference of civil engineering*, Semnan, Iran (in Persian).

29. Gharehbaghi, S.; Salajegheh, E. & Khatibinia, M. (2011). Optimization of reinforced concrete moment–resistant frames based on uniform hysteretic energy distribution, *proceeding of the 1st international conference on urban construction in the vicinity of active faults*, Tabriz, Iran.

30. Gharehbaghi, S. (2011). Damage and energy concepts-based optimum seismic design of RC structures. *M.Sc. Thesis*, Department of Civil Engineering, Shahid Bahonar University of Kerman, Iran (in Persian).

31. Gharehbaghi, S.; Salajegheh, E. & Khatibinia, M. (2012). Evaluation of seismic energy demand of reinforced concrete moment resistant frames considering soil–structure interaction effects, *proceedings of the eleventh international conference on computational structures technology*, Topping, B.H.V. (Ed.); Civil–Comp Press, Dubrovnik, Croatia.

32. Gharehbaghi, S. & Fadaee, M.J. (2012). Design optimization of RC frames under earthquake loads. *International Journal of Optimization in Civil Engineering*, IUST, Vol. 2(4), pp. 459–477.

33. Kaveh, A. & Zakian, P. (2014). Optimal seismic design of reinforced concrete shear wall-frame structures, *KSCE*

Journal of Civil Engineering, DOI 10.1007/s12205–014–0640–x, (in press).

34. Sexsmith, R.G. (1983). Bridge risk assessment and protective design for ship collision, *IABSE Colloquium,* Copenhagen, Denmark, pp. 425-433.

35. Coello Coello, C.A. (2002). Theoretical and numerical constraint-handling techniques used with evaluation algorithms: a survey of the state of art,*Computer Methods in Applied Mechanics and Engineering*, Vol. 191, pp. 1245-1287.

36. Rajeev, S. & Krishnamoorthy, C.S. (1992). Discrete optimization of structures using genetic algorithm,*Journal of Structural Engineering,* ASCE, Vol. 118(5), pp. 1233-1250.

37. Salajegheh, E.; Salajegheh, J.; Seyedpoor, S.M. & Khatibinia, M. (2009). Optimal design of geometrically nonlinear space trusses using adaptive neuro-fuzzy inference system, *Scientia Iranica*, Vol. 6(5), pp. 403-414.

38. Khatibinia, M. & Khosravi, Sh. (2014). A hybrid approach based on an improved gravitational search algorithm and orthogonal crossover for optimal shape design of concrete gravity dams, *Applied Soft Computing*, Vol. 16, pp. 223-233.

39. Khatibinia, M. & Naseralavi, S. S. (2014). Truss optimization on shape and sizing with frequency constraints based on orthogonal multi-gravitational search algorithm, *Sound and Vibration*, Vol. 333, pp. 6349–6369.

40. Shinozuka, M. & Sato, Y. (1967). Simulation of nonstationary random processes.*Journal of Engineering Mechanics,* ASCE, Vol. 93, pp. 11-40.

41. Mckay, M.D.; Beckman, R.J. & Conover, W.J. (1979). A comparison of three methods for selecting values on input variables in the analysis of output from a computer code, *Technometrics,* Vol. 21, pp. 439-445.

42. ACI318 Committee. (2005). *Building code requirements for structural concrete and commentary,* American Concrete Institute, Michigan, USA.

43. Kunnath, S.K.; Reinhorn, A.M. & Lobo, R.F. (1992). *IDARC Version 3: A program for the inelastic damage analysis of RC structures.* Technical Report NCEER-92-0022, National Center for Earthquake Engineering Research, State University of New York, Buffalo.

44. Park, Y.J. & Ang, A.H. (1985). Mechanistic seismic damage model for reinforced concrete, *Journal of Structural Engineering,* Vol. 111, pp. 722-39.

45. Moustafa, A. (2011). Damage-based design earthquake loads for SDOF inelastic structures, *Journal of Structural Engineering,* ASCE, Vol. 137, pp. 456–467.

46. Moustafa, A. & Mahmoud, S. (2014). Damage assessment of adjacent buildings under earthquake loads, *Engineering Structures,* Vol. 61, pp. 153–165.

47. Heidari, A. & Gharehbaghi, S. (2015). Seismic performance improvement of special truss moment frames using damage and energy concepts,*Earthquake Engineering and Structural Dynamics,* DOI: 10.1002/eqe.2499, (in press).

48. Gharehbaghi, S.; Salajegheh, E.; Shojaee, S. & Khatibinia, M. (2009). Optimum seismic design of RC structures considering damage index,*proceeding of the 1st international conference of lightweight structures and earthquake,* Kerman, Iran (In Persian).

49. Mazzoni, S.; McKenna, F.; Scott, M.H. & Fenves G.L., (2010). *OpenSEES: Open system for earthquake engineering simulation.* Pacific Earthquake Engineering Research Centre (PEER): http://opensees.berkeley.edu/.

50. Kent, D.C. & Park, R. (1997). Flexural members with confined concrete, *Structural Division,* ASCE, Vol. 97, pp. 1969-1990.

51. Saatcioglu, M. & Razvi, S.R. (1992). Strength and ductility of confined concrete,*Journal of Structural Engineering,* ASCE, Vol. 118, pp. 1590-1607.

52. Paulay, T. & Priestley, M.J.N. (1992). *Seismic design of reinforced concrete and masonry buildings*, 1st Ed. New York: Wiley-Interscience.

53. Zhang, Y.; Conte, J.P.; Yang, Z.; Elgamal, A.; Bielak, J. & Acero, G. (2008). Two-dimensional nonlinear earthquake response analysis of a bridge-foundation-ground system, *Earthquake Spectra*, Vol. 24(2), pp. 343-386.

54. Lysmer, J. & Kuhlemeyer, R.L. (1969). Finite element model for infinite media, *Journal of Engineering and Mechanics,* ASCE, Vol. 95, pp. 859-77.

55. Scott S.M. & Fenves G.L. (2010). Krylov subspace accelerated Newton algorithm: application to dynamic progressive collapse simulation of frames, *Journal of Structural Engineering*, ASCE, Vol. 136, pp. 473-80.

56. Clough, R.W. & Penzien, J. (1975). *Dynamics of Structures*, Mc Graw Hill.

57. Rousseeuw, P.J. & Leroy, A. (1987). *Robust regression and outlier detection*, New York: Wiley-Interscience.

58. Cawley, G.C. & Talbox, N.L.C. (2004). Fast exact leave-one-out cross-validation of sparse least squares support vector machinesm, *Neural Networks*; Vol. 17(10), pp. 1467–1475.

59. Cawley, G.C. (2006). Leave-one-out cross-validation based model selection criteria for weighted LS-SVM, *proceedings of the international joint conference on neural networks,* Vancouver, BC, Canada, pp. 1661–1668.

60. Seyedpoor, S.M.; Salajegheh, J.; Salajegheh, E. & Gholizadeh, S. (2009). Optimum shape design of arch dams for earthquake loading using a fuzzy inference system and wavelet neural networks, *Engineering Optimization,* Vol. 41(5), pp. 473-493.

61. Wu, Q. (2011). Hybrid model based on wavelet support vector machine and modified genetic algorithm penalizing Gaussian noises for power load forecasts, *Expert Systems with Applications,* Vol. 38, pp. 379-385.

62. Calisir, D. & Dogantekin, E. (2011). An automatic diabetes diagnosis system based on LDA-wavelet support vector machine classifier, *Expert Systems with Applications*, Vol. 38, pp. 8311-8315.

63. Zavar, M.; Rahati, S.; Akbarzadeh, M.R. & Ghasemifard, H. (2011). Evolutionary model selection in a wavelet-based support vector machine for automated seizure detection, *Expert Systems with Applications*, Vol. 38, pp. 10751-10758.

64. Zhang, L.; Zhou, W.; & Jiao, L. (2004). Wavelet support vector machine. IEEE Transactions on Systems, Man, and Cybernetics, Part B: Cybernetics; Vol. 34(1), pp. 34–39.

65. Khatibinia, M.; Salajegheh, E.; Salajegheh, J. & Fadaee M.J. (2012). Prediction of failure probability for soil-structure interaction system using modified ANFIS by hybrid of FCM-FPSO, *Asian Journal of Civil Engineering*, Vol. 13, pp. 1-27.

66. Bezdek, J.C. (1981). *Pattern Recognition with Fuzzy Objective Function Algorithms*, Plenum Press, New York.

67. Mehdizadeh, E.; Sadinezhad, S.; &Tavakkoli Moghaddam, R. (2008). Optimization of fuzzy clustering criteria by a hybrid of PSO and fuzzy c-means of fuzzy clustering algorithm, *Iranian Journal of Fuzzy Systems*, Vol. 5, pp. 1-14.

68. Abbas, AM and manohar CS (2007). Reliability-based vector nonstationary random critical earthquake excitations for parametrically excited systems, *Structural Safety*, Vol. 29, pp. 32-48.

6

CHAPTER

Geotechnical, Mineralogical and Chemical Characterization of the Missole II Clayey Materials of Douala Sub-Basin (Cameroon) for Construction Materials

Elisabeth Olivia Logmo[1], Gilbert François Ngon Ngon[1*], Williams Samba[1], Michel Bertrand Mbog[2], Jacques Etame[1]

[1]*Laboratory of Subsurface, Department of Earth Sciences, Faculty of Science, University of Douala, Douala, Cameroon*

[2]*Laboratory of Environmental Geology, Department of Earth Sciences, Faculty of Science, University of Dschang, Dschang, Cameroon*

ABSTRACT

Geotechnical tests conducted on clayey materials of Missole II, Douala sub-basin of Cameroon showed that these materials present: fines particles (55 to 78 wt.%), sand (22 to 44 wt.%), and plasticity index of 13.8 to 21.6%. The X-ray diffraction (XRD) and the chemical analysis revealed a kaolinite amount of 46 to 56 wt.%, 19 to 27 wt.% of illite, 12 to 19 wt.% of quartz, 3 to 5 wt.% of goethite, 2 to 5 wt.% of hematite, 1.5 to 5 wt.% of anatase, 2 to 3 wt.% of feldspar-K with 52.87 to 63.11 wt.% of SiO_2, 18.08 to 24.31 wt.% of Al_2O_3, 3.28 to 11.45 wt.% of Fe_2O_3 and a small content of bases (<2 wt.%). The results of geotechnical tests combined to those of the XRD and the chemical analysis showed that the Missole II clayey materials are suitable for the manufacture of bricks, tiles and sandstones.

Keywords: Cameroon; Construction Materials; Geotechnical Tests; Mineralogical and Chemical Analyses; Missole II

1. INTRODUCTION

Clay-rich materials are intensively used in the manufacturing of ceramics and as construction materials. However, the clayey deposit types are known as sedimentary, alluvial, and residual. A good knowledge of their occurrences, quantity and properties is required for their efficient exploitation.

In the Douala sedimentary sub-basin (South-Cameroon, Central Africa), some studies are done concerning the stratigraphic and tectonic evolution including [1-13]. Others studies are done to set up industrial units for manufacturing construction materials and ceramics [14] or concerning the mineralogical and chemical or thermal characteristics of the clay sediments [15-21].

In the Missole II area (Douala sub-basin), a geological study is carried out to locate and describe the clayey material outcrops and their provenance [22-24]. Also, some authors showed the possibility to obtain good ceramic building materials by mixing silica, feldspars, and kaolinitic and illitic clay of the Missole II area [25]. However, despite the preliminary works done on the field to describe clayey materials and to determine their sedimentation evolution, no further physical, mineralogical and chemical study is carried out to show their characteristics in order to their applications.

To that effect, the objective of this study is to associate the geotechnical characteristic to the mineralogical and chemical compositions of the clay occurrences of the Missole II deposit in order to evaluate its suitability for manufacturing of construction materials and ceramics.

2. GEOGRAPHIC AND GEOLOGICAL SETTING

Missole II is located on the Eastern part of the Douala sub-basin (Cameroon, Central Africa) between latitude 3°59' - 3°54'N and longitude 9°54' - 9°58'E. It is located within a humid equatorial climatic zone. Annual rainfall ranges between 3000 and 5000 mm, and the annual average temperature is 26°C. The vegetation is a dense rainforest transformed by human activities [26]. The geomorphology of the study area is a domain of the Cameroon coastal plain with low altitudes (40 - 120 m). The Missole II area shows hills with flat and sharp summits and is deeply dissected by V and U shaped valleys of MBongo, Bongougou, Missolo and Bongo the main rivers of the area. According to the geological map of SNH/UD report [12], the relative age of the Missole II sediments is Paleocene-Eocene corresponding to the N'Kapa Formation **Figure 1**.

The lithostratigraphy of Douala sub-basin is made of seven major Formations related to its geodynamic and sedimentary evolution [10-12]. 1) The synrift period represented by the Mundeck Formation (Aptian-Cenomanian) is discordant onto the Precambrian basement and consists of continental and fluvio-deltaic deposits, i.e., clays, coarse-grained sandstones, conglomerates. The postrift sequence includes; 2) the Logbadjeck Formation (Cenomanian-Campanian), discordant onto the Mundeck Formation and composed of micro conglomerates, sand, sandstone, limestone, and clay; 3) the Logbaba Formation (Maastrichtian), mainly composed of sandstone, sand and fossiliferous clay; 4) the N'kapa Formation (Paleocene-Eocene), rich in marl and clay with lenses of sand and fine to coarse-grained crumbly sandstone; 5) the Souellaba Formation (Oligocene) lying unconformably on N'kapa deposits and characterized by marl deposits with some interstratified lenses and sand channels; 6) the Matanda Formation (Miocene), dominated by deltaic facies interstratified with volcanoclasties layers; and 7) the Wouri Formation (Plio-Pleistocene) which consists of gravelly and sandy deposits with a clayey or kaolinic matrix.

3. RAW MATERIALS AND EXPERIMENTAL METHODS

The raw material used in this study comes from four representative clayey profiles of the Missole II area with three profiles of the interfluves along the Douala-Edéa road and one from the pit drilling on the lower slope of the valley. A geological survey shows different types of sediments with micro conglomerates, sandstones, fragments of ferruginous duricrusts and clays. Clayey layers are overlain upwards by sandstones or micro conglomerates, ferruginous duricrusts and sandy-clays, and occupy the lower part of the profiles. Four clayey facies identified with different mixed textures like sandy-clay, clayey-

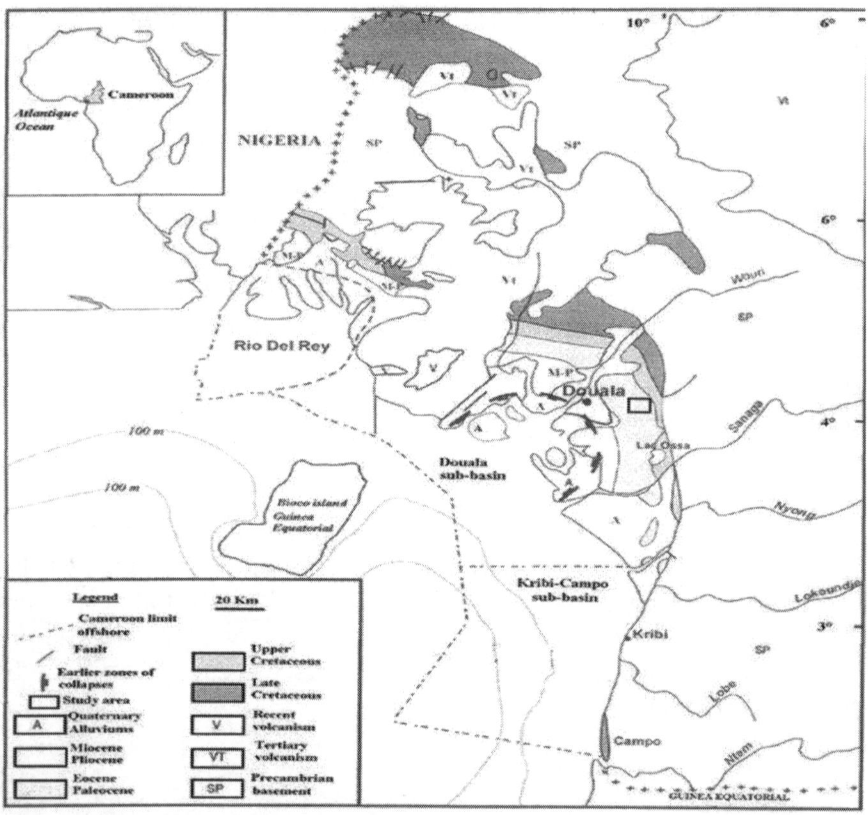

Figure 1. Geological sketch map of Cameroonian coastal basins (SNH/UD, 2005).

silt and silty-clay are mainly of sedimentary origin [23, 24]. The average thickness of the exploitable layers is 2.5 m. Two clayey samples collected from clay layer of each representative profile for mineralogical and chemical data are mixed to obtain average sample which served to realize the geotechnical analyses. A sufficient quantity of the single mixture of 2 to 3 kg of sediments is collected from a meter-long groove. The sample is analyzed for physical, mineralogy and chemistry.

The particle size distribution has been achieved in two steps: 1) a conventional sieving for the 63 to 2000 µm fractions; 2) using a sedigraph 5000 in automatic procedure for clay and silt fractions. The liquid limit is measured by the method of the dish of Casagrande (wL) and the plastic limit by the method of the roller (wP). The blue methylene value (Vb) is determined on the total sample.

One hundred grams of each homogenized sample is grounded to −200 mesh (0.075 mm) in an agate mortar for chemical and XRD mineralogical study. Mineral identification is performed using a Setsys 2400 apparatus from SETARAM 85 equipped with a DSC 1500 heat system with Pt crucibles for thermal analysis, from room temperature up to 1100°C using a rate treatment of $10°C{\cdot}min^{-1}$and alumina heat treated at 1500°C serving as reference material. For XRD, a Brünker diffractometer D8 ADVANCE with a copper source (λ = 1.5489 Å) is used on bulk and fine (<2 µm) samples, working under 40 kV and 40 mA. The exposure time for qualitative analysis is 2 h. Mineralogical phases are identified (JCPDS, 1998). Semi-quantitative analysis is performed [27]. For microscopic analysis, clay samples are examined with a scanning electron microscope (SEM) (Cambridge stereos can 200) coupled with an energy dispersive spectra microprobe (EDS). Homogenized powder of sediment sample is chemically analyzed for major and trace elements by ICP-AES after dissolution using acid digestion procedure with HF, HNO_3, and $HClO_4$. Classification of the clayey materials is performed by Autret (1983) method [28,29]. This method differentiates lateritic materials to non lateritic materials. It is based on the ratio S/R where, S is the ratio between SiO_2 concentration and the molar mass and R the ratio between Al_2O_3 plus Fe_2O_3 concentrations and their molar mass. In fact, for true lateritic material S/R is less than 1.33, for lateritic rocks it is 1.33 to 2, and for clay material it is more than 2.

4. RESULTS

4.1. Geotechnical Characteristics

The particle size distribution of the raw basic materials shows that the samples consist of 23 to 45 wt.% of sand, 17 to 33 wt.% of silts and 34 to 45 wt.% of clayey fractions. The geotechnical characteristics of the samples are reported in **Table 1**. These results are permitted to deduce that the class of these samples is in fine fractions [30]. The plasticity index (13.8% to 21.6%) is varied between 12 and 25, and shows that these materials are averagely plastic [31]. This plasticity is presumably a consequence of the average content of clayey minerals and quartz. The methylene blue value (0.4 to 0.91 g/100g) of the concerned materials, which is corroborated with the plasticity index, may suggest the absence of swelling clay minerals.

4.2. Mineralogical Composition

4.2.1. DTA analysis

Figure 2 represents the DTA analysis curves obtained for the raw materials M2P3 and M2A3 at room temperature up to 1100°C using a rate treatment of 10°C·min^{-1}. Differential scanning calorimetric (DSC) observed on these DTA analysis curves shows endothermic pheno menon towards 110°C associated to the physisorbed water and another endothermic phenomenon towards 320°C due to the dehydroxylation of goethite. A specific endothermic transformation at 400°C - 600°C is due to the loss of structural hydroxyl groups of the kaolinite and an exothermic peak at 900°C - 1000°C due to the structural reorganization of the metakaolinite [32,33].

Table 1. Geotechnical characteristics of the samples.

Geotechnical characteristics	Results			
Average samples	**M2A2**	**M2P3**	**M2P4**	**M2A3**
Depth (cm)	600 - 750	700 - 800	700 - 800	50 - 150
Particles size distribution (wt.%)				
Sand (2000 - 63 µm)	45	31.3	44	22
Silts (63 - 2 µm)	21	25.5	17	33
Clays (<2 µm)	34	43.2	39	45
Atterberg limits				
Liquid limit, wL (%)	35.11	40.87	32.9	47.45
Plasticity limit, wP (%)	19.84	22.92	19.07	25.81
Plasticity index, PI (%)	15.27	17.95	13.83	21.64
Blue methylene value of the total sample (g/100g)	0.4	0.82	0.7	0.91

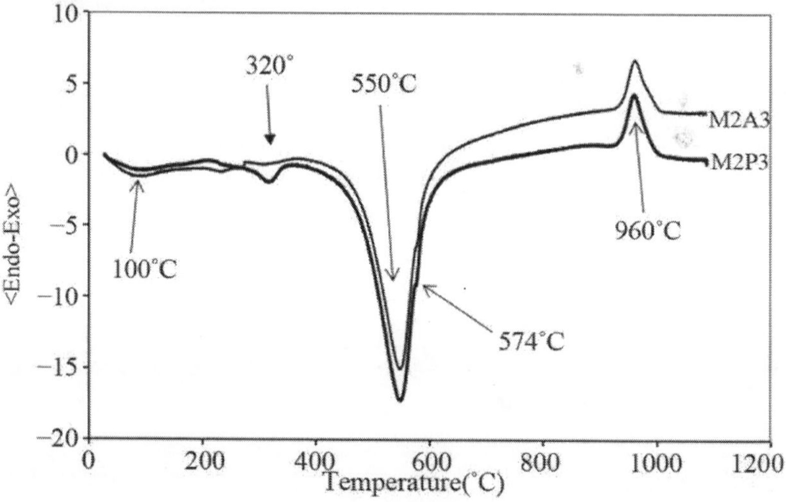

Figure 2. DTA curves of M2A3 and M2P3 materials for a heating rate of 10°C/min.

4.2.2. X-Ray Diffraction

The diffractograms of powder of the bulk samples ground until <80 µm revealed that clayey materials are kaolinitic and illitic and have permitted the detection of the following mineral phases: kaolinite, quartz, illite, goethite, hematite and accessory feldspar-K and anatase [24]. These mineral phases were usually present in theclayey materials of the Douala sub-basin [16,17,21].

The diffractograms of <2 µm fraction of **Figure 3** also indicated the presence of the same mineral phases observed in the total samples with kaolinite, illite, quartz, goethite, hematite and accessory feldspar-K and anatase. Comparison of the minerals detected, these mineral phases are present in the fraction of the smaller and thicker granulometry.

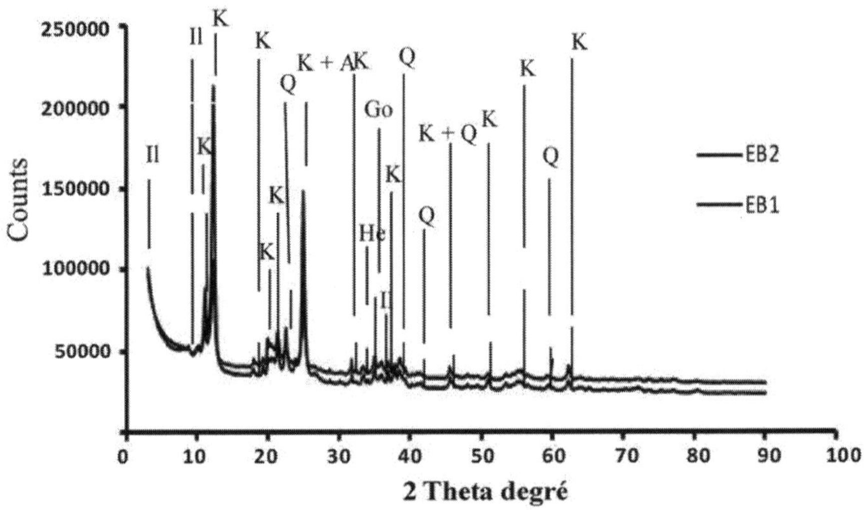

(a)

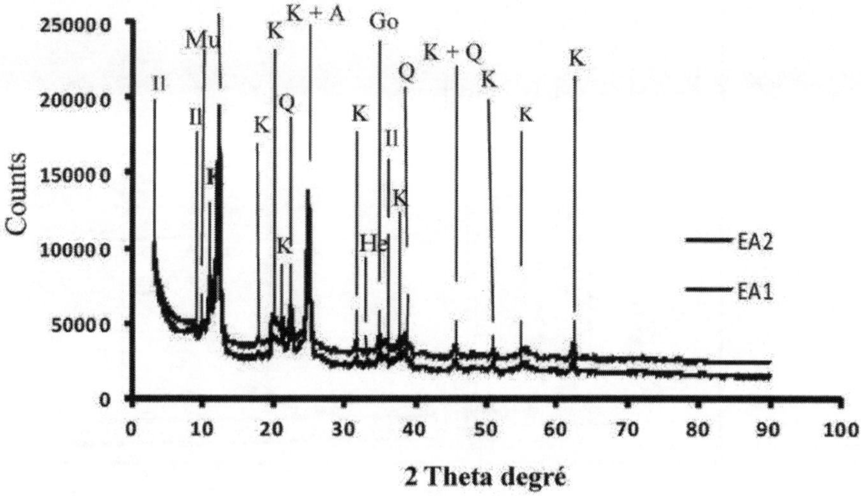
Counts
2 Theta degré
EA2
EA1
Il
Mu
K
K + A
Go
K + Q
Q
He
0
50000
100000
150000
200000
250000
0
10
20
30
40
50
60
70
80
90
100

(b)

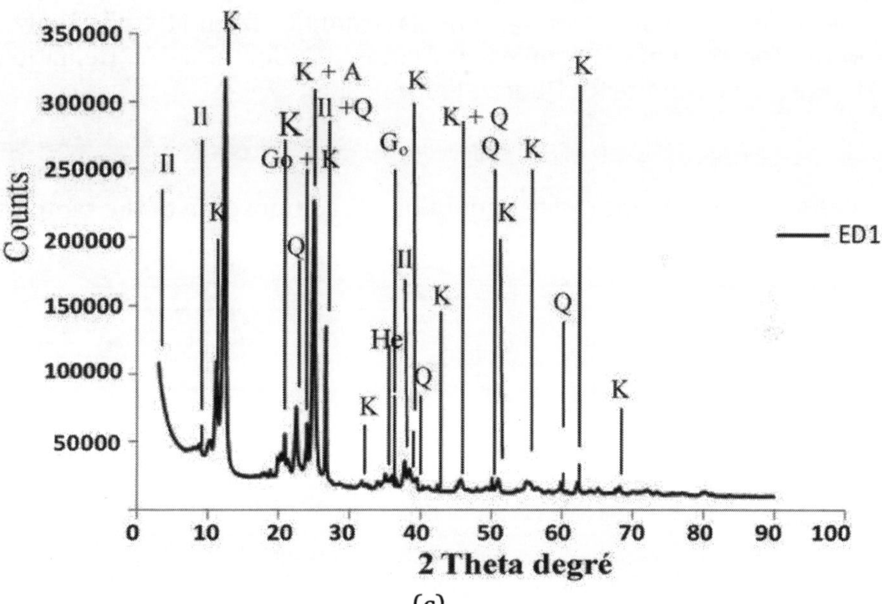
Counts
2 Theta degré
ED1
Il
K
K + A
Il +Q
Go + K
Go
K + Q
Q
He
50000
100000
150000
200000
250000
300000
350000
0
10
20
30
40
50
60
70
80
90
100

(c)

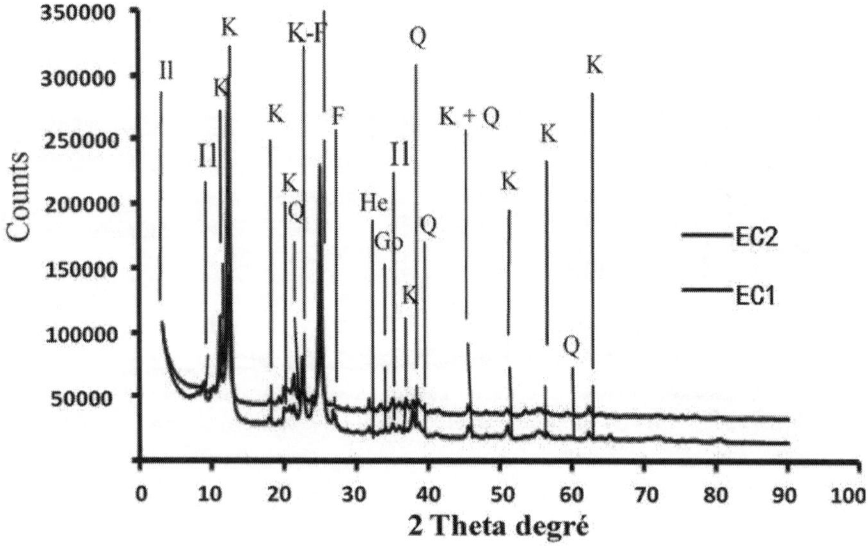

(d)

Figure 3. XRD patterns of <2 μm clay samples from Missolle II area of (a), (b), (c) and (d) profiles. A: Anatase; G: Goethite; He: Hematite; Il: Illite; K: Kaolinite; Q: Quartz; F: K-feldspar.

Table 2. Semi-quantitative mineralogical composition of the samples (wt.%).

	Profiles of the interfluves									Profiles of the pit	
Profiles	a			b			c			d	
Samples	EA1	EA2	M2A2	EB1	EB2	M2P3	EC1	EC2	M2P4	ED1	M2A3
Depth (cm)	600 - 700	700 - 750	600 - 750	700 - 750	700 - 750	750 - 800	700 - 750	750 - 800	700 - 800	50 - 100	50 - 150
Kaolinite	53.9	52.1	53.6	46.4	50.4	48.1	47.7	48.5	48.5	47.8	48.4
Illite	23.5	20.5	20.8	26.3	19.9	23.4	26.6	19.2	20.8	25.0	23.6
Quartz	16.1	15.0	17.1	18.8	17.7	18.2	12.5	17	16.9	13.7	16.1
Goethite	3.0	4.0	4.2	3.9	3.0	4.0	3.2	4.4	3.8	5.0	4.1
Hematite	-	2.2	1.6	-	3.0	2.1	3.2	4.4	3.6	2.9	3.2
Anatase	3.5	3.6	3.3	4.6	-	3.3	4.3	1.5	3.3	2.9	3.1
Feldspath-K	-	-	-	-	3.0	1.2	-	2.2	1.6	-	-

The **Table 2** gives the semi-quantitative mineralogical composition of the mineral phases of the samples. It reported an important proportion of kaolinite (46 to 56 wt.%), appreciable amounts of illite (19 to 27 wt.%) and quartz (12 to 19 wt.%), which are in relation with the value of the plasticity index (13.8 to 21.6%) and the methylene blue value (0.8 to 1.6 g/100g). The amounts of the other mineral phases are respectively goethite (3 to 5 wt.%), hematite (2 to 5 wt.%), anatase (1.5 to 5 wt.%) and feldspar-K (2 to 3 wt.%).

4.2.3. Microscopic Analysis

Figure 4 shows a SEM observation performed on samples of grey and mottled, with sandy-clay, silty-clay and clayey-silt texture. In clayey materials, piles of disordered kaolinites are observed with various sizes and irregular forms, which indicated that kaolinites of sedimentary clayey material of the Missole II area are poorly crystallized [34].

4.3. Chemical Composition

The chemical composition of the samples is given on **Table 3**. The results of the chemical analysis shows the presence of important amounts of SiO_2 (52.87 to 63.11 wt.%), associated with an appreciable amounts of Al_2O_3 (18.08 to 24.31 wt.%) and Fe_2O_3 (3.28 to 11.45 wt.%) and a small content of bases (<2 wt.%, CaO + MgO + Na_2O + K_2O).

These results show that quartz, alumino-silicates com pounds and iron minerals are predominate in the samples. The weak content of oxide calcium or potassium has proven that the calcitic and potassic minerals like calcite or feldspar are absent or are in a weak proportion. With S/R more than 2, the analyzed samples are clays [28].

M2P3

M2A3

Figure 4. SEM images of M2P3 and M2A3 clayey materials of Missole II.

Table 3. Chemical composition of the sample.

							Profiles of the interfluves					Profiles of the pit
Profiles			a			b			c		d	
Samples	DL	EA1	EA2	M2A2	EB1	EB2	M2P3	EC1	EC2	M2P4	ED1	M2A3
Depth (cm)		600 - 700	700 - 750	600 - 750	700 - 750	700 - 750	750 - 800	700 - 750	750 - 800	700 - 800	50 - 100	50 - 150
SiO_2	0.01	63.11	56.85	58.01	54.97	57.24	56.23	52.87	54.02	54.08	63.1	62.84
Al_2O_3	0.01	18.94	24.31	23.06	22.16	19.1	20.87	23.24	20.59	21.01	18.08	18.61
Fe_2O_3	0.01	3.28	3.45	3.39	5.96	7.54	8.02	3.51	11.45	8.89	4.0	3.54
CaO	0.01	0.04	0.02	0.02	0.02	0.2	0.02	0.02	0.02	0.02	0.02	0.02
MgO	0.01	0.1	0.1	0.1	0.13	0.08	0.1	0.13	0.08	0.12	0.1	0.1
Na_2O	0.01	0.06	0.06	0.06	0.1	0.03	0.1	0.06	0.06	0.06	0.06	0.06
K_2O	0.01	0.43	0.4	0.49	0.77	0.21	0.3	0.47	0.47	0.47	0.6	0.7
TiO_2	0.01	1.02	1.38	1.42	0.99	1.1	1.0	1.2	1.2	1.2	0.78	0.83
MnO	0.01	0.01	0.01	0.01	0.01	0.01	0.01	0.01	0.01	0.01	0.01	0.01
P_2O_5	0.01	0.02	0.04	0.04	0.04	0.04	0.03	0.02	0.05	0.05	0.03	0.03
Cr_2O_3	0.01	0.01	0.01	0.01	0.01	0.02	0.01	0.01	0.02	0.01	0.01	0.01
LOI		13.02	13.12	13.29	14.21	13.85	13.21	17.38	11.81	14.01	14.4	12.34
Total		99.94	99.75	99.9	99.37	99.22	99.9	99.29	99.79	99.93	99.25	99.34

DL: Detection limit; LOI: Loss on ignition.

5. DISCUSSION

The particle size in this study revealed texture classes: silty-clay, clayey-silt and sandy-clay described in many deposits of the Douala sub-basin [17,21]. These classes characterized very sticky and plastic materials in accordance with some field characteristics such as smoothness, stickiness and plasticity [35]. Indeed, the fine particles and especially the very fine clay minerals determine the plasticity of an earthly material [32]. The plasticity index (13.8% to 21.6%) of the above clayey samples is supported by the high amounts of fine particles (silt + clay) and fine clay minerals. The amount of clay fraction is used to determine the manufactured products as is shown in **Table 4**[36]. However, particle size and plasticity index are amongst the main geotechnical characteristics used to determine a convenient choice of construction material. In this case, taking into account of the amounts of sand (22 to 44 wt.%), fine particles notably the clay fractions (34 to 45 wt.%), and plasticity index, the Missole II clayey materials are convenient to the manufacturing of bricks, tiles and sandstones [37-39].

The mineralogical composition and geotechnical characteristics of a construction material must be known simultaneously in order to make a suitable choice for this raw basic material. It is well known that clay minerals have detrimental effects on the geotechnical properties; since some swelling clay minerals such as smectites absorb more water than the other (e.g. kaolinite). The volumetric shrinkage which increases with swelling clay minerals such as smectites constitutes an important parameter to take into account for civil public engineering [29,40]. The amount of clay minerals and especially swelling clay minerals contribute to the increasing of volumetric shrinkage and the swelling phenomena. Some authors joined the swelling potential and/or volumetric shrinkage to two parameters [36]. A first ordering is based on the plasticity index and the percentage of fine clayey fraction (<2 μm size). A second method proposes to use the liquid limit and the percentage of fine particles which size is less than 74 μm [41]. In this case, the studied materials, which have plasticity index between 12% and 25%, 34 and 45 wt.% of clayey fraction (<2 μm size), 54 and 78 wt% of fine fraction (<80 μm size) and liquid limit between 32% and 47% are very low to moderate swelling materials as is indicated in **Table 5**. This observation is corroborated with the mineralogical composition which shows that fine fraction is mainly composed of kaolinite and few illites, as well as smectites are absent.

Table 4. Relation between particles size and type of ceramic product to realize (after Cere and Mazel, 1993).

ϕ < 2 μm (wt.%)	Ceramic products to realize
5 - 25	Full bricks
25 - 35	Perforated bricks
35 - 45	Hollow bricks, drains
40 - 50	Tiles, hordes

The geotechnical and mechanical characteristics of clayey materials are essentially depending on their chemical and mineralogical compositions as well as of the minerals distribution [37,39]. In this study, the clayey samples mainly constituted of poorly crystallized kaolinite (46 to 56 wt.%), which is in agreement with the loss on ignition (12% to 17%), illite (19 to 27 wt.%) and quartz (12 to 19 wt.%), are adequate for the composition of ceramic pastry [42] and for the manufacturing of bricks and tiles [14,37]. Also, clayey materials are siliceous (SiO_2 > 60 wt.%), with their alumina (Al_2O_3) content less than 35 wt.% and appreciable amount of iron (<10 wt.% with the exception of EC2 sample) is suitable for the manufacturing of tiles and bricks [41]. However, the weakest amounts of bases <2 wt.% (oxides of calcium, magnesium, sodium and potassium) which are the smelting or cimentitious compound corroborated with the flexural strength less than 10 MPa of the Missole II clayey material [25].

Table 5. Swelling potential of soils according to their plasticity index, liquid limit and respectively their <2 and <74 µm fractions.

PI (%)	% <2 µm	% <74 µm	wL (%)	Swelling potential
>35	>95	>95	>60	Very high
22 - 35	60 - 95	60 - 95	40 - 60	High
18 - 22	30 - 60	30 - 60	30 - 40	Moderate
<18	<30	<30	<30	Low

6. CONCLUSIONS

The following conclusions can be drawn out from the present work:

1) The geotechnical, mineralogical and chemical results show that the clayey materials of the Missole II, Douala sub-basin of

Cameroon are mainly constituted of sand (23 to 45 wt.%), silts (17 to 33 wt.%) and clays (34 to 45 wt.%) with plasticity index of 13.8% to 21.6%.

2) Kaolinite, illite and quartz are the main minerals.

3) Clayey materials are also siliceous and aluminous, and have appreciable amount of iron in most clayey samples (Fe_2O_3 less than 10 wt%).

4) Based on the obtained geotechnical, mineralogical and chemical results, and the literature data, the studied materials are suitable for the manufacturing of bricks, tiles and sandstones.

5) These clayey materials present poor mechanical characteristics (feeble flexural strength and great fragility) partly linked to the chemical and mineralogical nature fundamentally complex of natural clays.

7. ACKNOWLEDGEMENTS

The authors acknowledge the financial support of the University of Douala and the geochemical analysis of Geo Labs (Geoscience Laboratories) in Canada. Many thanks are also given to Professor David Smith, Director of the GEMH laboratory at ENSCI and Professor Gisèle Lecomte of Centre Européen de la Céramique, Limoges, France, and to all of their collaborators.

REFERENCES

1. Y. C. Belmonte, “Stratigraphie du bassin sédimentaire du Cameroun,” Proceeding on 2nd West African Micropaleontology Colloquium, Ibadan, 1-5 September 1966, pp. 7–24.

2. M. E. Brownfield and R. R. Charpentier, “Geology and Total Petroleum Systems of the West-Central Coastal Province (7203) West Africa,” USGS: Geological Survey Bulletin, 2006, 2207-B 52 p.

3. E. Dartevelle and P. Brebion, "Mollusques fossiles du Crétacé de la Côte occidentale d'Afrique du Cameroun à l'Angola,"Annales du Musée Royal Congo Belge, Sciences Géologiques de Tervuren I-Gastéropodes, Vol. 8, No. 20, 1956, pp. 1-128.

4. E. Dartevelle, S. Freinex and J. Sornay, "Mollusques fossiles du Crétacé de la Côte occidentale d'Afrique du Cameroun à l'Angola,"Annales du Musée Royal Congo Belge, Sciences Géologiques de Tervuren II-Lamellibranches, Vol. 8, No. 20, 1957, pp. 1-271.

5. J. F. Dumort, "Identification par la Telédétection de l'Accident de la Sanaga (Cameroun)," Géodynamique, Vol. 1, No. 3, 1968, pp. 13-19.

6. J. B. Meyers, B. R. Rosendahl and H. Groschel-Becker, "Deep Penetrating MCS Imaging of the Rift-to-Drift Transition Offshore Douala and North Gabon Basins West Africa," Marine Petrology Geology, Vol. 13, No. 7, 1996, pp. 791-835.

7. D. Reyre, "Histoire géologique du bassin de Douala," In: D. Reyre, Ed., Symposium sur les Bassins Sédimentaires du Littoral Africain, Association du Service Géologique d'Afrique, IUGS, 1966, pp. 143-161.

8. P. R. N. Ngaha, "Contribution à l'Etude Géologique, Stratigraphique et Structurale de la Bordure du Bassin Atlantique du Cameroun," Thèse 3e Cycle, Université de Yaoundé, 1984.

9. P. Maurizot, A. Abessolo, J. L. Feybesse, V. Johana and P. Lecomte, "Synthèse des Travaux de 1978 a` 1986," Rapport 85CM066, 1986.

10. J. M. Regnoult, "Synthèse Géologique du Cameroun," D.M.G., Yaoundé, Cameroun, 1986.

11. F. R. Nguene, S. Tamfu, J. P. Loule and C. Ngassa, "Paleoenvironments of the Douala and Kribi/Campo Subbasins in Cameroon, West African," Colloque de Géologie Africaine, Libreville, Recueil des Communications, Géologie Africaine, 6-8 May 1991, 1992, pp. 129-139.

12. SNH/UD, "Stratigraphie Séquentielle et Tectonique des Dépôts Mésozoïques Synrifts du Bassin de Kribi/Campo," Rapport Non Publié, 2005.

13. C. S. Manga, "Stratigraphy, Structure and Prospectivity of the Southern Onshore Douala Basin Cameroon—Central Africa," In: M. J. Ntamak-Nida, G. E. Ekodeck and M. Guiraud, Eds., Cameroon and Neighboring Basins in the Gulf of Guinea (Petroleum Geology tectonics Geophysics Paleontology and Hydrogeology), African Geoscience, 2008, pp. 13-37.

14. P. M. Thibaut and P. Le Berre, "Recherche d'Argiles Pour Briques Dans la Région de Yaoundé, Douala et Edéa," Rapport 85CM065, 1985.

15. D. Njopwouo, "Minéralogie et Physico-Chimie des Argiles de Bomkoul et de Balengou (Cameroun). Utilisation Dans la Polymérisation du Styrène et Dans le Renforcement du Caoutchouc Naturel," Thèse d'Etat, Faculté des Sciences, Université de Yaoundé, 1984.

16. D. Njopwouo and R. Wandji, "Minéralogie de l'Argile Kaolinique de Bomkoul (Cameroun)," Revue de Sciences et Technique, Série des Sciences de la Terre, I 3-4, 1985, pp. 71-81.

17. D. Njopwouo and S. Kong, "Minéralogie de la Fraction Fine des Matériaux Argileux de Bomkoul et de Balengou (Cameroun)," Annales de la Faculté des Sciences, Série des Sciences Chimiques, I 1-2, 1986, pp. 17-31.

18. A. Elimbi and D. Njopwouo, "Firing Characteristics of Ceramics from the Bomkoul Kaolinite Clay Deposit (Cameroon)," Tile and Brick International, Vol. 18, No. 6, 2002, pp. 364-369.

19. A. M. M. Mpondo, "Cartographie des Affleurements de la Localité de Bomkoul (Sous-Bassin de Douala)," Université de Douala, Mémoire, 2010.

20. M. B. Mbog, "Etude Morphologique, Physico-Chimique et Minéralogique des Argiles de Bomkoul Dans le Sous-Bassin

Sédimentaire de Douala-Cameroun," Université de Douala, Douala, 2010.

21. G. F. Ngon Ngon, J. Etame, M. J. Ntamak-Nida, M. B. Mbog, A. M. Maliengoue Mpondo, M. Gérard, R. Yongue-Fouateu and P. Bilong, "Geological Study of Sedimentary Clayey Materials of the Bomkoul Area in the Douala Region (Douala Sub-Basin, Cameroon) for the Ceramic Industry," Comptes Rendus Geoscience, Vol. 344, No. 6, 2012, pp. 366-376.

22. W. Samba, "Etude Morphologique, Géotechnique et Minéralogique des Argiles de Missole 2 Dans le SousBassin de Douala-Cameroun," Université de Douala, Douala, 2010.

23. E. O. Logmo, "Etude Géologique, Minéralogique et Géochimique des Argiles de Missole II Dans le Sous-Bassin de Douala (Cameroun)," Université de Douala, Douala, 2012.

24. G. F. Ngon Ngon, E. Bayiga, M. J. Ntamak-Nida, J. Etame, S. Noa Tang and R. Yongeu-Fouateu, "Trace Elements Geochemistry of Clay Deposits of Missole II from the Douala Sub-Basin in Cameroon (Central Africa): A Provenance Study," Sciences, Technologie et Développement, Vol. 13, No. 1, 2012, pp. 20-35.

25. G. F. Ngon Ngon, G. L. Lecomte Nana, R. Yongue Fouateu, G. Lecomte and P. Bilong, "Physicochemical and Mechanical Characterisation of Ceramic Materials Obtained from a Mixture of Silica, Feldspars and Clay Material of the Douala Region in Cameroon (Central Africa)," Advances in Ceramic Science and Engineering (ACSE), Vol., 2, No. 1, 2013, pp. 23-31. [Paper reference:2]

26. R. Letouzey, "Atlas du Cameroun, Phytogéographie Camerounaise," Imprimerie Nationale Yaoundé, 1968.

27. A. K. Chakravorty and D. K. Ghosh, "Kaolinite-Mullite Reaction Series: The Development and Significance of a Binary Aluminosilicate Phase," Journal of the American Ceramic Society, Vol. 74, No. 6, 1991, pp. 1401-1406.

28. P. Autret, "Latérites et Graveleux Latéritiques," Laboratoire Central des Ponts et Chaussées, 1983.

29. M. Younoussa, "Etude Géotechnique, Chimique et Minéralogique de Matières Premières Argileuses et Latéritiques du Burkina Faso Améliorées aux Liants Hydrauliques: Application au Génie Civil (Bâtiment et Route)," Thèse, Université de Ouagadougou, 2008.

30. NF P94-057, "Analyse Granulométrique des Sols. Méthode par Sédimentation," AFNOR, 1992.

31. NF P94-051, "Détermination des Limites d'Atterberg," AFNOR, 1993. [Paper reference:1]

32. C. A. Jouenne, "Traité de Céramiques et Matériaux Mineraux," Septima, Paris, 1990.

33. S. Lee, Y. J. Kim and H. S. Moon, "Phase Transformation Sequence from Kaolinite to Mullite Investigated by an Energy-Filtering Transmission Electron Microscope," Journal of the American Ceramic Society, Vol. 82, No. 10, 1999, pp. 2841-2848.

34. M. W. Carty, "The Colloidal Nature of Kaolinite," The Bulletin of the American Ceramic Society, Vol. 78, 1999, pp. 72-76. [Paper reference:1]

35. E. A. Fitzpatrick, "Soils-Their Formation, Classification and Distribution," Longman, Berlin, 1983.

36. L. Cere and F. Mazel, "Caractérisation d'Argiles," ENSCI Limoges, 1993.

37. Y. Beron and P. Le Berre, "Guide de Prospection des Matériaux," Manuels et Methodes du BRGM, No. 5, 1983.

38. V. Rigassi, "Bloc de Terre Comprimée," Manuel de Prospection, Vol. 1, Craterre EAG, 1995.

39. N. Française, "Sols: Reconnaissance et Essais. Description-Identification-Dénomination des Sols," XP P94-011, 1999.

40. A. A. Al-Rawas and M. Qamarouddin, "Construction Problems of Engineering Structures Founded on Expansive Soils and Rocks in Northern Oman," Building and Environment, Vol. 33, No. 2-3, 1998, pp. 159-171.

41. A. Djedid, A. Bekkouche and A. M. Aissa Mamoune, "Identification and Prediction of the Swelling Behavior of Some Soils from the Tlemcen Region of Algeria," Bulletin des Laboratoires des Ponts et Chaussées, Vol. 233, 2001, pp. 69-77.

42. O. Castelein, "Influence de la Vitesse du Traitement Thermique sur le Comportement d'un Kaolin: Application au Frottage Rapide," Thèse, Université de Limoges, 2000.

7

CHAPTER

Real-Time Monitoring of Climactic and Geotechnical Variables during Landslides on the Slopes of Serra do Mar and Serra da Mantiqueira (São Paulo State, Brazil)

Rodolfo Moreda Mendes[1], Mário Valério Filho[2]

[1]*National Center for Monitoring and Warning of Natural Disasters-CEMADEN, São José dos Campos, Brazil*

[2]*IP & D, Vale do Paraíba University, São José dos Campos, Brazil*

ABSTRACT

The municipalities of Ubatuba, Campos do Jordão, and São José dos Campos are located in the region of São Paulo State (Brazil). These municipalities are recognized nationally for having an elevated number of recorded landslides on slopes and embankments. In addition, these municipalities contain multiple areas that are at risk for landslides. Various soil landslides occurred in these municipalities in January 2013, when real-time climactic and geotechnical variables were monitored by automatic rain gauges, humidity sensors and soil temperature and suction devices. The resulting data were used to understand the functions of each variable in the occurrence of land- slides. Analyses of rainfall, humidity and soil temperature were used with field investigations to formulate a hypothesis regarding the predominant rupture mechanism and the role of each monitored variable in the deflagration of the soil landslides that occurred in the three studied municipalities. The geotechnical variable data revealed that both temperature and soil moisture contents played fundamental roles in the deflagration of shallow planar landslides in urban areas. The hourly rain intensity and/or rainfall accumulation for 24 and/or 72 h were responsible for the deflagration of the landslides that occurred in the studied areas, along with the existing anthropic constraints in the risk areas. Significant variations did not occur in the soil suction data during the landslides, principally due to the unsatisfactory sensor precision when reading field suction between −10 and −100 kPA (±25%).

Keywords: Landslides, Urban Area, Real-Time Monitoring, Analysis of Rainfall and Geotechnical Parameters

1. INTRODUCTION

Geologic risks in São Paulo State (Brazil) have been highlighted nationally and internationally, mainly due to the elevated number of landslides that have occurred on urban slopes and the number of municipalities with areas of landslide risk. Several municipalities that are located in regions with elevated landslide susceptibility are notable [1] . For example, the municipalities with urban areas on the slopes of the Serra da Mantiqueira (Campos do Jordão and São José dos Campos) and Serra do Mar (Ubatuba). The susceptibility of these areas to landslides has been corroborated by the significant number of landslides that have been registered by the Geological Institute of São Paulo State (Figure 1) in the framework of the Preventative Plan of Civil Defense (PPDC).

The identification, evaluation and management of landslide risk areas in São Paulo State (Brazil) began following large scale accidents with grave consequences that occurred during the summer of 1987/1988 in the region of Serra do Mar. Consequently, the state government decided to implement the PPDC, which was specific for landslides on the slopes of the Serra do Mar in Sao Paulo State. The main objective of the PPDC is to avoid human death by relocating populations that occupy risk areas before a landslide occurs. However, to prevent death from landslides, the actions of the PPDC are based on the daily monitoring of critical rainfall levels (rain accumulated in 72 h) and the meteorological forecasts for each municipality. The critical rainfall levels used in the preventative removal action of populations from risk areas were defined based on a correlative study between rainfall and landslides that was initially designed for the municipality of Cubatão, São Paulo State [2] [3] .

The correlation results obtained in [2] [3] for the municipality of Cubatão were posteriorly extrapolated for the remaining municipalities of São Paulo State by using the 72-h accumulated rainfall data as a reference. With this method, different values of critical rain accumulation were established for the municipalities using the PPDC. These values highlighted the 72-h rainfall accumulations in the following regions: Serra do Mar (Ubatuba)—120 mm; Vale do Paraiba (São José dos Campos)—100 mm and Serra da Mantiqueira (Campos do Jordão)—80 mm.

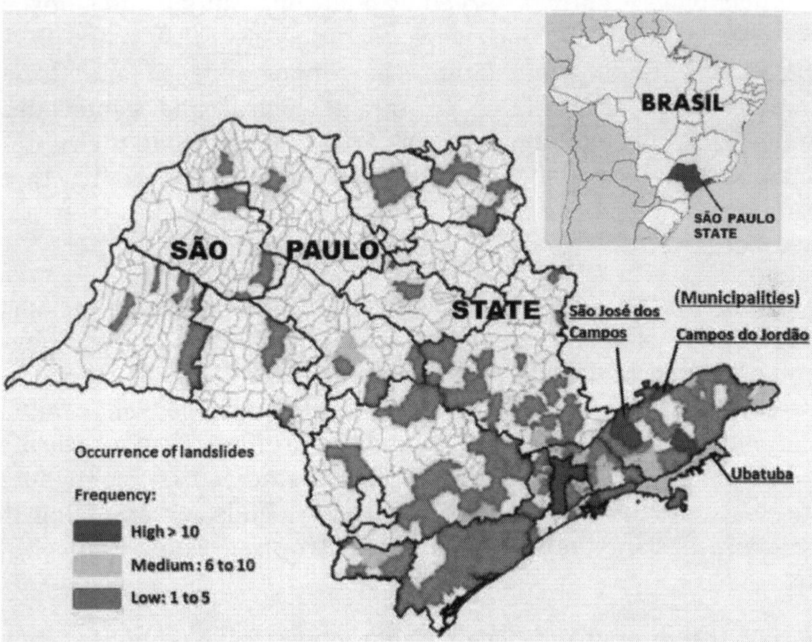

Figure 1. Spatial distribution of the occurrences of landslides in the municipalities of the São Paulo State in the period 2000 to 2010 (Source: Geological Institute).

A significant number of national and international studies were observed to establish correlations between rainfall and gravitational mass movements. Among these international studies, developments in [4] -[20] are notable. In Brazil, the studies of [2] [3] [21] -[31] are noteworthy.

However, the spatial distribution of rainfall in a particular region is not homogeneous and tends to be less correlated with landslides when rain gauges are farther apart, especially when convective rainfall occurs in Brazil during the summer season, which generally has high spatial variability [32] . Therefore, the extrapolation of rainfall data points that represent a given area where landslides have occurred represents critical rainfall values in other regions that may contain inadmissible uncertainties for the operation of the PPDC. In addition, the critical rainfall rates used in certain areas cannot satisfactorily represent the critical rainfall conditions that are necessary for the deflagration of landslides in other regions that have different geological and geotechnical characteristics. This disconnect was recently observed by analyzing rainfall and landslide data within the PPDC that were recorded over 10 years [33] . Overall, this finding highlights the importance of monitoring parameters that are related to the geological and geotechnical characteristics of each region or risk area to understand the rupture mechanisms of landslides in urban areas, especially those parameters that permit infiltration and rainwater retention assessments.

Studies of climate and geotechnical variables on slopes in the Serra do Mar, specifically on the northern coast of São Paulo (Brazil), have been conducted by Brazilian researchers since 2005. These studies have contributed to understanding the effects of seasonal rainfall variations on the suction and humidity profiles of non-saturated residual soils [34] -[38] . Moreover, these studies have contributed to the use of sensors and non-invasive methods for assessing the suction profiles and moisture contents of tropical residual soils [38] - [43] .

In Brazil, the climate and peculiar characteristics of tropical residual soils have generated the need for specific studies that combine climactic variables and soil geotechnical properties to understand the hydro-mechanical behavior of the soil and the relationships between climate and the geotechnical variables that trigger landslides in such soils.

Residual soil landslides are often associated with the rainy season on Brazilian slopes. According to [44] , the main geotechnical variable for the occurrence of landslides is the reduction or elimination of shear soil strength that results from rainwater infiltration in the soil profile. For landslides in tropical residual soils, it is important to study the seasonal variations of matrix suction and moisture content within the soil profile that result from periods of rain and drought. Thus, it is important to evaluate the influence of rainfall on the hydro-mechanical behavior of the soil by monitoring climactic aspects and their effects on the matrix suction and moisture contents of the soil profile. Therefore, this paper presents new real-time monitoring results from geotechnical and climactic variables that are obtained by installing automatic rain gauges and sensors. These sensors measured total rainfall, suction, temperature and soil moisture in the municipalities of Campos do Jordão, São José dos Campos and Ubatuba, São Paulo State (Brazil), where several landslides occurred in risk areas during the in situ monitoring period.

2. STUDY AREAS

2.1. Selection and Location

The initial criteria for selecting the study areas for implementing the field trials considered the following regional aspects of the physical and urban environment of each municipality: 1) distinct regional climate variations (in terms of rainfall distribution), 2) regional aspects of relief and 3) the presence of landslide risk areas. In addition, the following conditions were used to choose the study areas: 1) areas with high and/or very high risk levels for landslides, 2) areas with safe areas for equipment installation and monitoring and 3) areas with adequate mobile phone signal to enable data transmission via GSM/GPRS technology.

The selected study areas are located in municipalities with individual climate features that are based on regional climates and reliefs (according to Martinelli 2010: Faixa Costeira (Ubatuba),

Porção Lestedo Planalto Atlântico (Campos do Jordão), and Vale do Paraíba (São José dos Campos). The locations of the experimental plots in the municipalities of Campos do Jordão (Ef-1), São José dos Campos (Ef-2) and Ubatuba (Ef-3) are shown in Figure 2.

Consequently, the regional climate and relief patterns for each municipality provide distinct average annual rainfall values. For example, Sera do Mar and Mantiqueira, where the municipalities of Ubatuba (Ef-3) and Campos do Jordão (Ef-1) are situated, have average annual rainfalls of between 1700 - 3000 mm, respectively. In addition, the average rainfall in Vale do Paraíba, where the municipality of São José dos Campos (Ef-2) is situated, is between 1200 and 1400 mm (Table 1).

2.2. Geological and Geotechnical Characterization

The geological and geotechnical characterization of each experimental region was initially performed in the field by observing the soil profiles that were considered the most critical (exhibiting greater natural susceptibility to landslides). These areas represented landslide risk areas (dashed yellow line) in the municipalities of Campos do Jordão (Ef-1), São José dos Campos (Ef-2) and Ubatuba (Ef-3) (Figure 3). Briefly, the following regional and local geological features of each experimental field were obtained:

Ø Campos do Jordão (Ef-1)—Metamorphic rocks, such as migmatites, gneisses, metadioritic and granulites dominate the region. In addition, schists, quartzites and metaconglomerates occur as lenses with some mag- matic intrusions [46] . Migmatitic rocks occur locally with alternating bands of neosome (granite) and paleosome (gneiss), with thicknesses that range from a few meters (in the case of granite) to centimeters (in the case of gneisses) thick [47] [48] .

Ø São José dos Campos (Ef-2)—The study area is regionally inserted into the tectonic compartment named the Southeast Regional Fold with rocks of the Embu Complex (paragneiss units and banded gneiss) that are composed of phyllites, schists, gneisses and migmatites. Locally rocks consisting of gneisses and migmatites are dominant without the marked occurrence of lithological variations [49] .

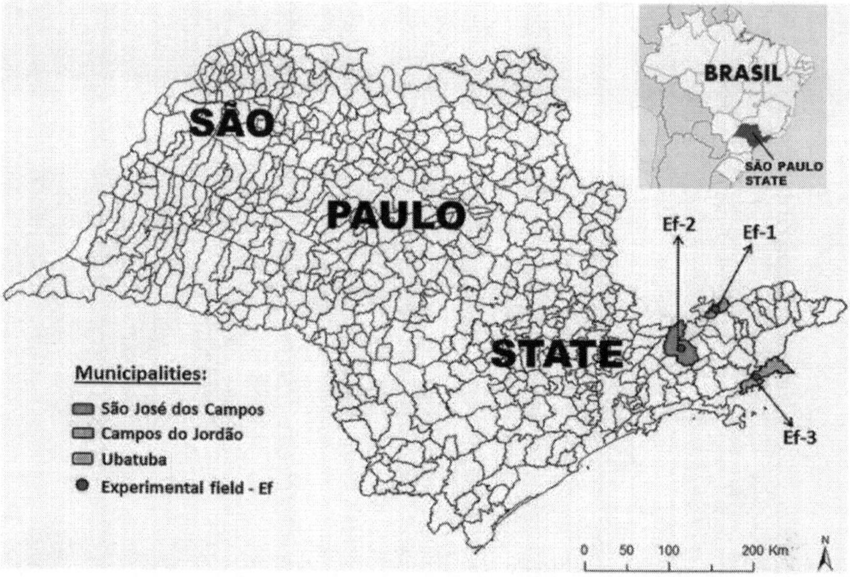

Figure 2. Localization of the municipalities in the São Paulo State (Campos do Jordão, São José dos Campos e Ubatuba) where were implanted the experimental fields (Ef-1, Ef-2 e Ef-3), respectively.

Table 1. Minimum, maximum and average rainfall annual values of the studied areas

Municipalities	Rainfall average annual (mm)			Period	Rain gauge localization		Altitude
	Minimum	Maximum	Average	(Year)	Latitude	Longitude	(Meters)
São José dos Campos	917	2.055	1.289	1943 to 2003	23°11'S	45°53'W	560
	887	2.271	1.396	1943 to 2000	23°03'S	45°54'W	570
Campos do Jordão	1.035	4.457	1.801	1937 to 2004	22°43'S	45°34'W	1590
	1.342	2.786	1.778	1973 to 2001	22°42'S	45°29'W	1620
Ubatuba	1.777	4.374	3.057	1957 to 2004	23°23'S	45°07'W	142
	1.414	3.977	2.399	1945 to 2000	23°23'S	44°50'W	184

The altitude is relative to sea level. Source of the data [45].

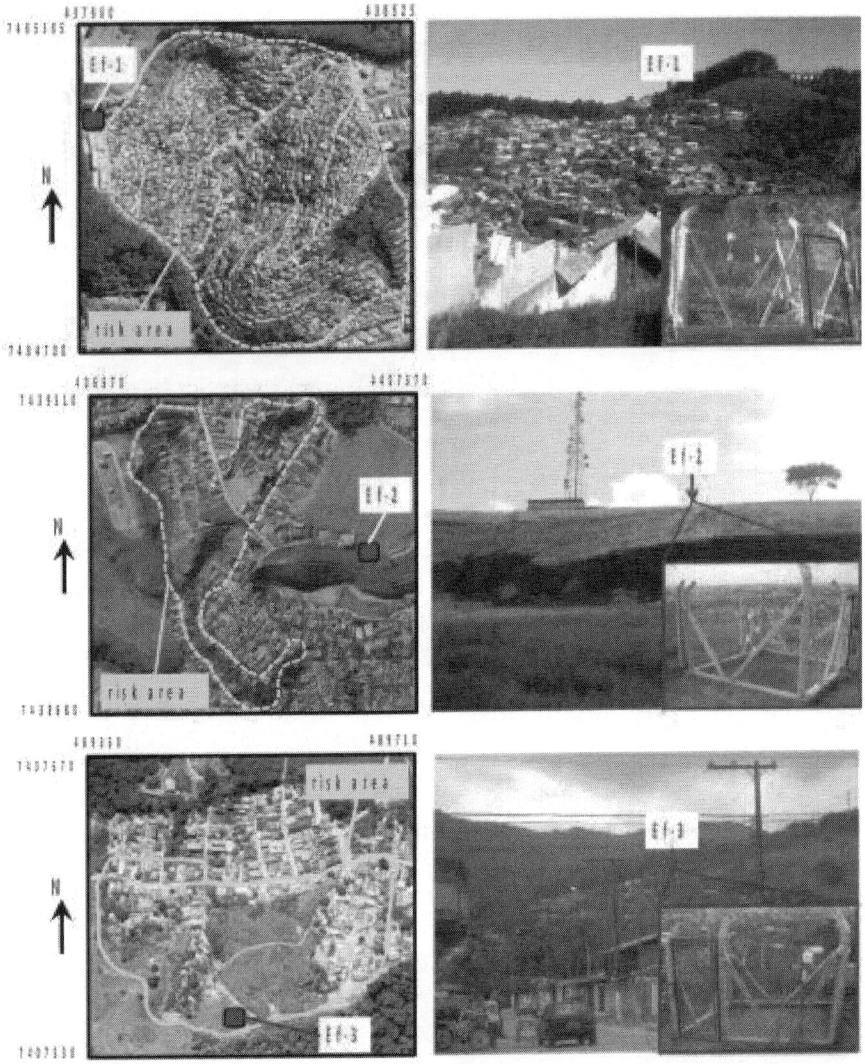

Figure 3. Localization of the experimental fields implanted in the landslides risk areas in the municipalities of Campos do Jordão (Ef-1), São José dos Campos (Ef-2) e Ubatuba (Ef-3).

Ø Ubatuba (Ef-3)—regionally occurring rocks grouped into five predominant lithologic assemblages of migmatites with diverse structures, granite-gneiss, biotite gneiss, quartzite and gneiss and gneisses changing to migmatite [50] [51] . Locally, rocks composed of porphyritic biotite granite and charnockites are not dominant.

Geotechnical characterization began with the identification of representative soil profiles in the experimental fields Ef-1, Ef-2 and Ef-3. These profiles had the following soil horizons: 1) Horizon "H1", topsoil resulting from pedogenesis with a thickness of 0.7 to 1.0 m; 2) Horizon "H2", the mature residual soil (no matrix rock structure and relatively homogeneous) with a thickness of 2.5 to 4.0 m; and 3) Horizon "H3", young residual soil or saprolites resulting from a rocky basement with the original matrix rock structure and a thickness of 5 to 15 m.

The thickness variations of the soil horizons were directly related to the position of the slope profile and its morphology, with greater depths at end regions and shallower portions at the bases of the slopes. The water table was considerably deeper on the slopes where experimental fields were planted and where upwelling or resurgences of water were not observed during field investigations. In Figure 4, schematic soil profiles of the three experimental fields are presented. These profiles were determined at the locations where the climactic and geotechnical monitoring systems were implemented.

After identifying the soil profiles of each experimental site, representative samples of each soil horizon (H1, H2 and H3) were collected for geotechnical characterization according to the Brazilian Technical Standards (ABNT). The geotechnical characterizations of sieve analysis, soil grain density, natural density and the Atterberg limits (liquid and plastic limits) were performed with deformed samples that were collected in the study sites of Campos do Jordão, São José dos Campos and Ubatuba. The overall results of the geotechnical characterization tests are shown in Table 2.

2.3. Landslides Characteristics in the Study Areas

Landslides in regions with the soils that often occur in the Serra do Mar and Serra da Mantiqueira (where the experimental sites are located) are usually associated with the rainy season [52] . Soil landslides, especially those of translational or planar shallow soils, occur naturally in middle and high slopes with rugged reliefs, even if the slopes are free of human influence.

The primary mechanism of slope destabilization is the reduction or elimination of "apparent" soil cohesion, which results from the infiltration of rainwater through the soil profile [22] [52] [53] . Because this type of destabilization is the dominant form of instability in the studied areas, it is important to monitor soil water infiltra- tion and retention (rainfall, moisture content, suction and temperature) continuously to understand the function of each variable in this type of instability mechanism.

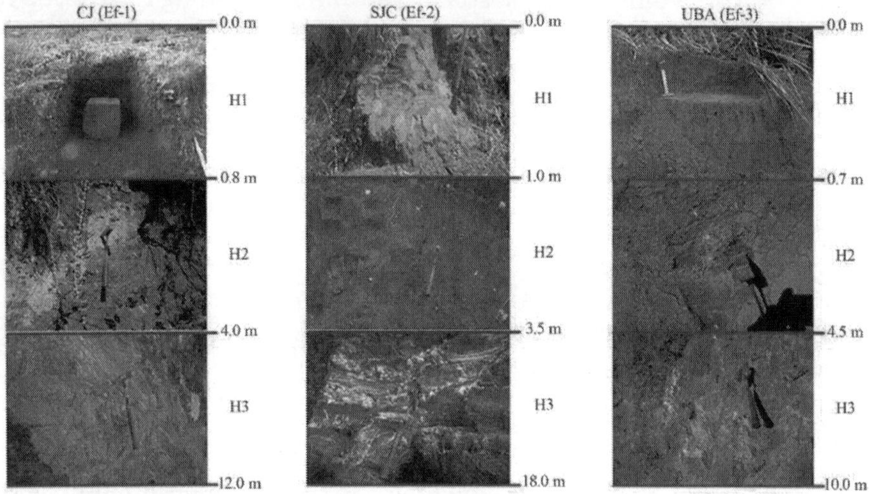

Figure 4. Soil profiles representative of the experimental fields of Campos do Jordão (Ef-1), São José dos Campos (Ef-2) e Ubatuba (Ef-3).

In general, the major natural constraints that are responsible for translational landslides in the study areas include high slope steepness (generally larger than 30°), which is associated with morphology (concave and/or linear geometry), and the presence of seasonal "apparent" cohesion, which results from an unsaturated soil profile and high rainfall (cumulative and/or hourly rainfall intensity). In addition, human action in these locations, such as slope cuts and/or dumped fills, overloading the land and the release of wastewater and/or leaking pipes on the slopes, significantly increases the geotechnical instability.

Table 2. Geotechnical characterization of the Ef-1, Ef-2 and Ef-3 experimental field.

Municipalities/ experimental field	Localization		Sample/ layer in soil profile	Depth sample (m)	Grain size (%)			Atterberg limits				γ_s	γ
	Long.	Lat.			Sand	Silt	Clay	w_L (%)	w_P (%)	IP (%)	USCS	(kN/m³)	(kN/m³)
Campos do Jordão (Ef-1)	45°36'W	22°44'S	CJ-H1	0.3	54	19	27	25	NP	NP	ML	29.6	13.1
			CJ-H2	3.0	42	14	44	31	29	2	ML	23.5	16.2
			CJ-H3	6.0	67	26	7	24	21	3	SM	27.0	13.8
São José dos Campos (Ef-2)	45°54'W	23°09'S	SJC-H1	0.5	56	18	26	29	NP	NP	SM	26.8	15.0
			SJC-H2	3.0	36	8	56	27	20	7	CL-ML	26.6	14.4
			SJC-H3	6.5	57	34	9	27	24	3	SM	26.9	15.7
Ubatuba (Ef-3)	45°06'W	23°26'S	UBA-H1	0.4	53	16	31	50	37	13	MH-ML	27.1	17.4
			UBA-H2	1.5	41	13	46	48	23	25	CL	26.3	16.5
			UBA-H3	7.0	54	20	26	45	26	19	CL	26.7	15.2

The landslides recorded in the Campos do Jordão (Ef-1) and São José dos Campos (Ef-2) study areas were planar or shallow translational (depth of rupture < 2.0 m) and occurred in urban areas. In this case, the three (03) following predominant rupture mechanisms were observed (Figure 5): A—surface rupture in the layer of residual soil slope H1; B—surface rupture in the residual soil layer H2 or H1 cut slope; and C—surface rupture at the base of the created fill or at the slope of the residual soil layer of H1 (with mobilization of the overlying dumped fill). Generally, the type "C" rupture mechanism moves greater amounts of material, which reach a larger number of homes and cause serious damage. The failure mechanisms of landslides that occurred in the city of Campos do Jordão between 01/11/2013 and 01/13/2013 are presented in Figure 5 (A, B, or C). Figure 6 shows the locations of the eight (08) landslides that occurred in the city of Campos do Jordão near the experimental field (Ef-1) between 01/11/2013 and 01/13/2013 (blue dots). In addition, Figure 6shows eighteen (18) landslides that were recorded in 2000 (red polygons on map).

Landslides that occurred in the city of São José dos Campos between 01/10/2013 and 01/11/2013 are presented in Figure 7. Figure 8 shows the locations of the two (02) landslides that occurred in the city of São José dos Campos near the experimental site (Ef-2) between 01/10/2013 and 01/11/2013 (blue dots). In addition, Fig. 8 shows the 61 landslides that were recorded between 2009 and 2013 (Figure 8, A).

The landslides that occurred near the experimental monitoring site of Campos do Jordão (Ef-1) and São José dos Campos (Ef-2) were affected by anthropic components, including instability factors. These factors included the implementation of slope cuts and dump fills, the discharge of wastewater and rainwater on slopes and/or slope cuts/landfills and the discharge of leaking pipes on slope or slope cuts/landfills. Consequently, rainfall, depending on its hourly intensity or total accumulation over 24, 48 or 72 h, was responsible for the outbreak of occasional landslides in these urban areas.

A landslide of soil and rock across a 14 km stretch of highway BR 101 is shown in Figure 9. This landslide occurred between January 3 and 4, 2013. Resurgences of water occurred in adjacent areas along the base of the slope based on the massive slow movement (crawling), the scattered erosion in the form of grooves and the presence of small to large blocks of rock at the base of the slope from the last event. This landslide process on the natural slope in the Serra do Mar was shallow planar and moved the residual soil layers of H1 and H2 at the top of the slope. The major predisposing factor of instability was the occurrence of the previous event, which removed a significant amount of material (soil and rock) at the base of the slope. This event was associated with heavy rain from January 3 to 4, 2013, based on the accumulated precipitation data that were recorded by the automatic environmental station of Picinguaba (near the landslide) (Figure 10).

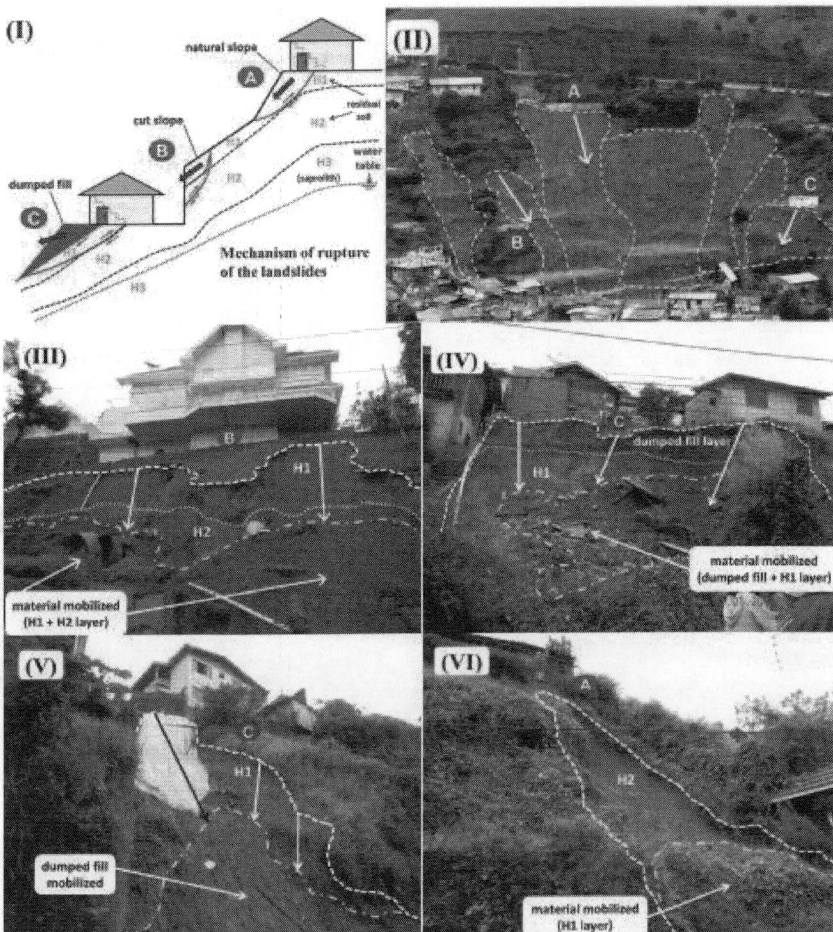

Figure 5. (I) Characteristics and main failure mechanisms of landslides in the study areas; (II) Landslides occurred in Campos do Jordão in the year 2000 [48] ; (III), (IV), (V) and (VI) landslides of the types B, C (H1), C (dumped fill) and A, respectively, occurred in Campos do Jordão between 01/11/2013 and 01/13/2013 (Source: Geological Institute).

Overall, three landslides occurred in Ubatuba between 01/03/2013 and 01/04/2013 (Figure 10; yellow dots 1, 2 and 3). These landslides occurred on natural, un-developed slopes. However, some type of intervention occurred at the base of the slopes (work projects cut/dumped fills for construction of highway BR 101).

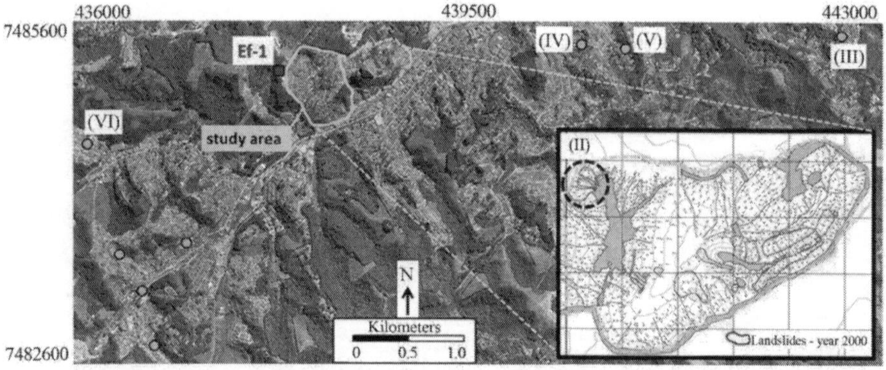

Figure 6. Landslides occurred in Campos do Jordão in the year 2000 (red polygons on map, [48]) and landslides occurred in Campos do Jordão between 01/11/2013 and 01/13/2013 (blue dots, source: Geological Institute), the numbers II, III, IV, V and VI refer to the landslides presented in Figure 5.

Figure 7. (VII) and (VIII) landslides of the type C (dumped fill) and B, respectively, occurred in São José dos Campos between 01/10/2013 and 01/11/2013 (Sources: VII, authors and VIII, Civil Defense of São José dos Campos).

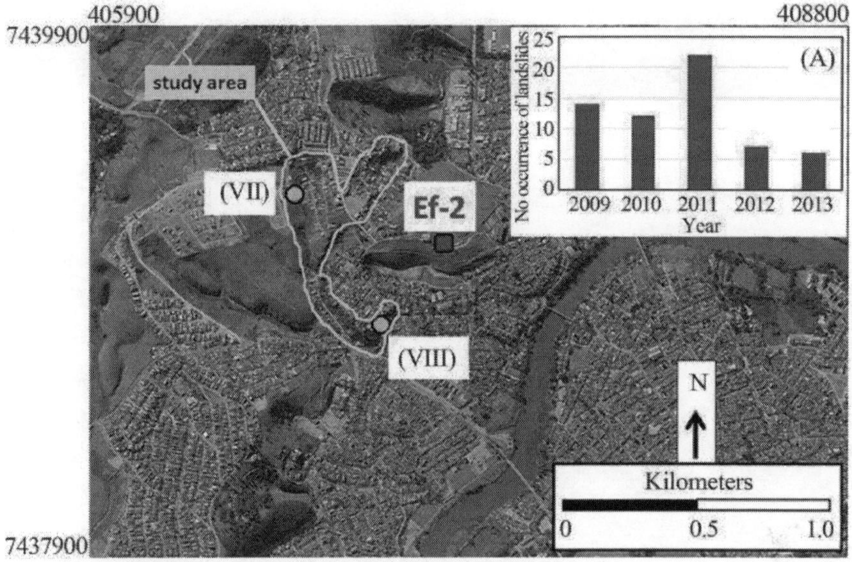

Figure 8. Localization of the landslides occurred in São José dos Campos between 01/10/2013 and 01/11/2013 (blue dots, sources: VII, authors, VIII, Civil Defense of São José dos Campos), the blue dots VII and VIII refer to the landslides presented in the Figure 7; (A) record of landslides occurred in São José dos Campos in the last five years (Source: Civil Defense of São José dos Campos).

During these events in Ubatuba, the automatic environmental station (AES/INPE) near the three landslide points (yellow dots 1, 2 and 3) recorded the following information between 01/03/2013 and 01/04/2013: 125.0 accumulated rainfall in 24 h, mean soil temperature of 28.5°C and a volumetric moisture content of 0.45 m^3/m^3 (Figure 10). Despite the considerable distance between the AES/INPE and the experimental field monitoring station of Ubatuba (Ef-3), very similar soil temperature (28°C) and moisture (0.46 m^3/m^3) data were recorded between the landslides (Figure 14). In addition, the amounts of accumulated precipitation over 24 h were significantly different (40.0 mm).

Figure 9. Reactivation of landslide on the slope of Serra do Mar (yellow dot 2 in Figure 10), occurred in Ubatuba between 01/03/2013 and 01/04/2013 in the km 14 of the highway BR-101 (Source: Geological Institute).

In general, the anthropic activities interfered with the landslide rupture mechanisms in the studied urban areas, especially for slope cuts and dumped fill materials, regarding the amounts of hourly or daily rainfall needed to promote landslides. Thus, human interventions in densely populated risk areas significantly reduce the amount of rain that is required to generate landslides relative to other areas with low levels of human occupation and intervention.

3. EXPERIMENTAL SITE IMPLEMENTATION

3.1. In Situ Monitoring Sensor Installation

Equipment and monitoring sensors were installed in soil profiles that represented the experimental soil profiles

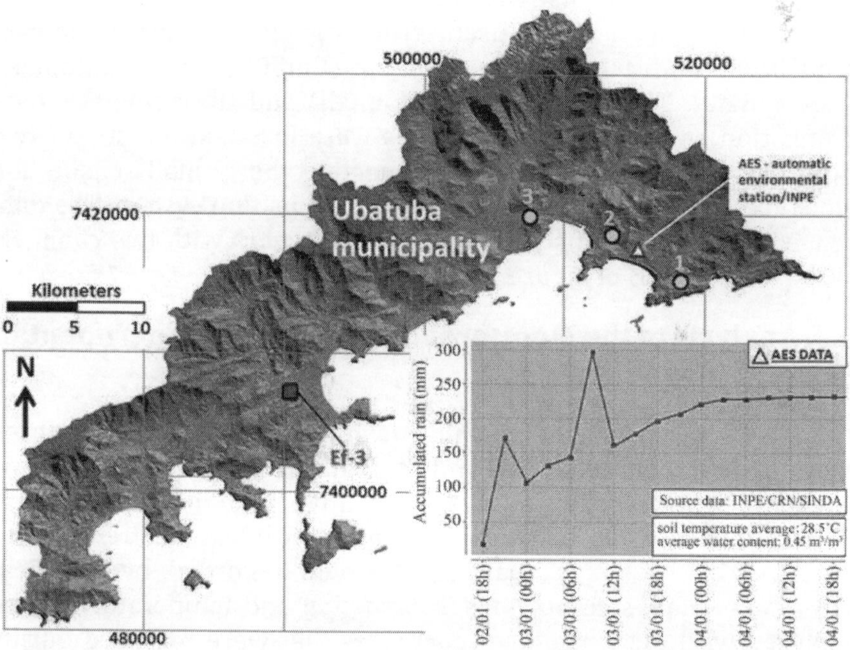

Figure 10. Localization of the landslides occurred in Ubatuba between 01/03/2013 and 01/04/2013 (yellow dots 1, 2, e 3, [54]), the yellow dot 2 refer to the landslide presented in Figure 9; the data of temperature and moisture of the soil represent the average values of the soil profile up 0.50 meter deep [55] .

in the Campos do Jordão (Ef-1), São José dos Campos (Ef-2) and Ubatuba (Ef-3) fields as shown inFigure 4. To install sensors and other equipment that make up the "Monitoring System of the Experimental Plots", the installation recommendations of the manufacturer (Decagon Devices) were followed. Briefly, the following procedures were used. 1) Soil suction and temperature sensor (MPS-2 dielectric potential sensor). Initially, an auger was used to make a hole that was of the depth of the sensor installation depth. Next, the sensor was placed in the "installation system" and wrapped in a moist mass of the same soil that was removed from the bottom of the hole. Finally, the sensor was inserted into the hole to its base (steps A/B/C of Figure 11). 2) Soil moisture sensor (10HS soil moisture). Initially, an auger was used to make a hole that was the depth of the sensor installation depth. Next, the sensor was inserted directly into the hole to its base. In this case, the "installation system" that was developed specifically for this project (steps D/E Figure 11) was used. 3) Rain gauge and automatic data acquisition and transmission system (Em50G collect automatic system date). The rainfall gauge (rainfall) and the automatic data acquisition and transmission devices were installed directly in two rigid PVC tubes before directly connecting the rainfall sensor, soil moisture and suction cables to the data acquisition system. The data acquisition system consisted of two data loggers with five channels each (steps F/G/H of Figure 11).

3.2. Analysis of the Monitored Data from the Experimental Plots

The in situ monitoring of climatic and geotechnical data in the three experimental plots (Ef-1, Ef-2 and Ef-3) allowed for the joint analysis of emergency records and the Preventive Plan of Civil Defense (PPDC) during a period with multiple landslides (12/01/2012 to 04/30/2013). For this analysis, the data recorded by the field sensors (rainfall, soil moisture, soil suction and temperature) were analyzed with the landslide occurrences that were recorded during the PPDC operation. The in situ measurements of rainfall, soil moisture, soil suction and temperature were taken at 10-minute intervals and the data were automatically stored in data loggers every 4 h for a total of six (06) measurements per day. In Figures 12-14, the data recorded by all of the monitoring

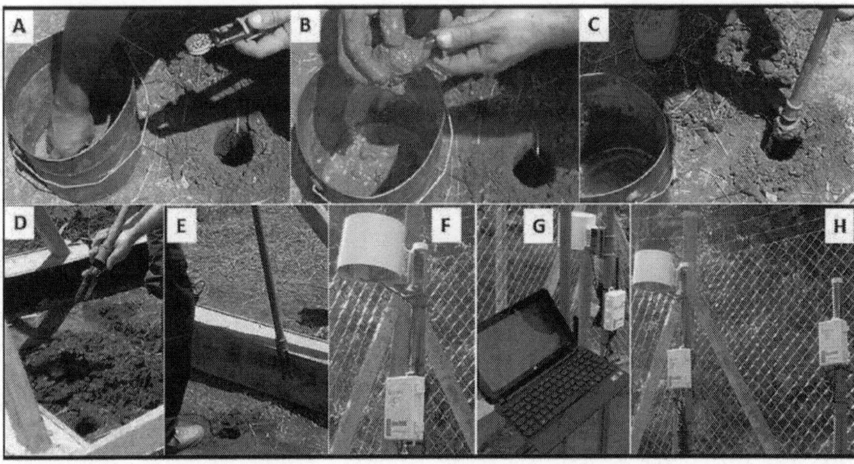

Figure 11. Installation of in situ monitoring sensors (A/B/C, soil suction and temperature sensors; D/E, soil moisture sensor), automatic data acquisition systems and rain gauges (F/G/H, rain sensor with data loggers with data transmission via the Global System for Mobile Communications, GSM/GPRS cellular technology).

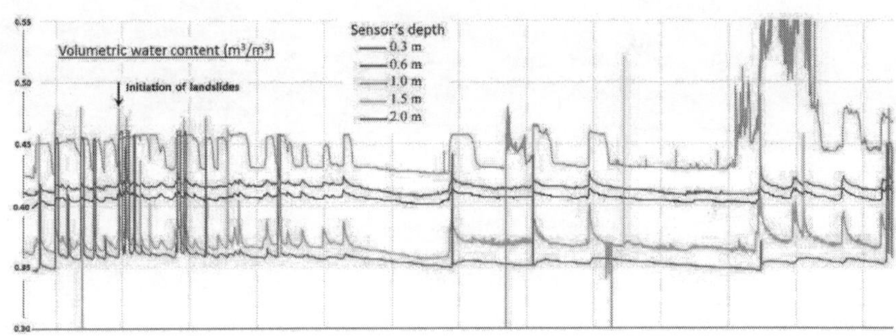

(a)

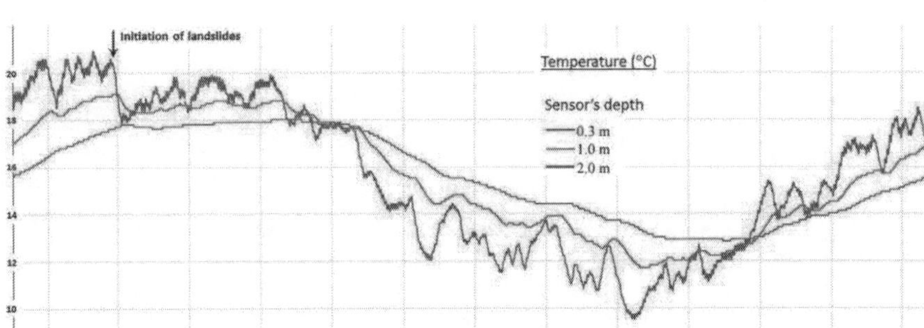

(b)

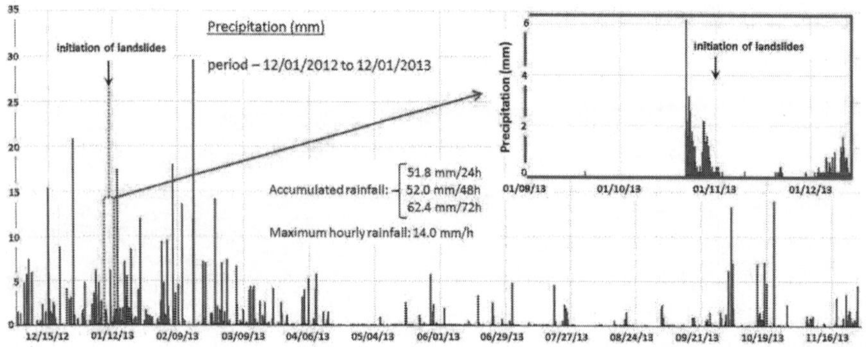

(c)

Figure 12. Monitoring data from the precipitation, soil temperature and soil moisture sensors in the experimental plot of Campos do Jordão (Ef-1) from 12/01/2012 to 12/01/2013, the main period of observed landslides (01/11/2013 to 01/12/2013, indicated by black arrows).

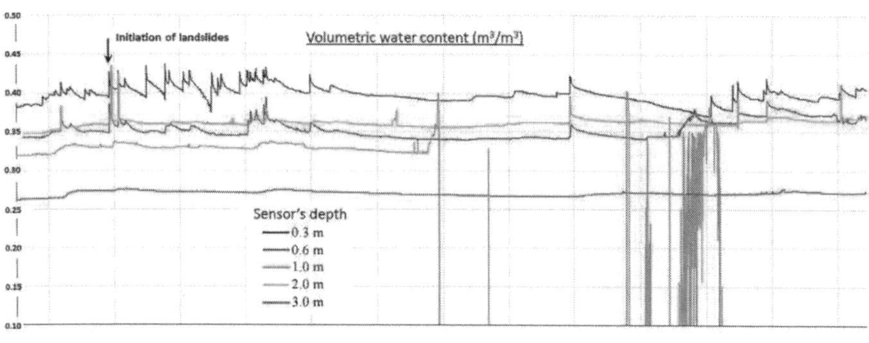

(a)

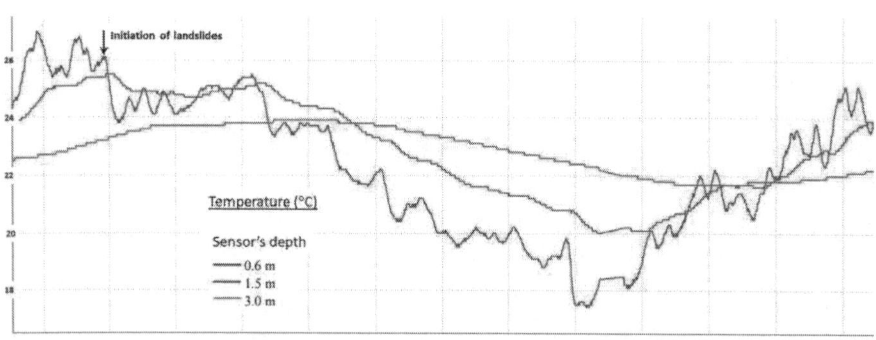

(b)

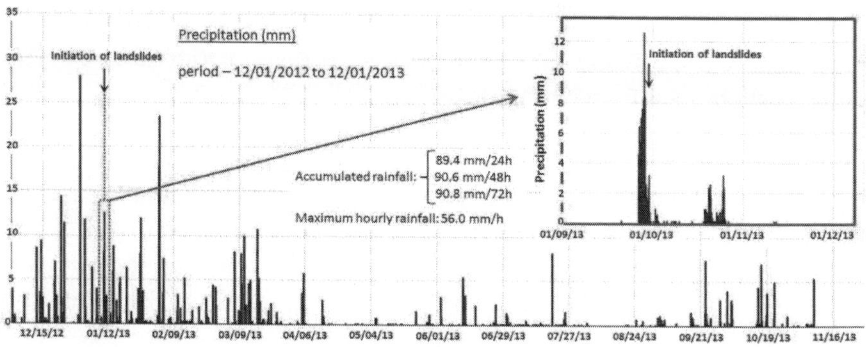

(c)

Figure 13. Monitoring data from the precipitation, soil temperature and soil moisture sensors in the experimental plot of São José dos Campos (Ef-2) from 12/01/2012 to 12/01/2013, the main period of observed landslides (01/10/2013 to 01/11/2013, indicated by black arrows).

systems in the three experimental fields are presented (Ef-1, Ef-2 and Ef-3) for the period between 12/01/2012 and 12/01/2013. Along with the climatic and geotechnical data that were recorded during the period, Figures 12-14 show the most probable regions of landslide outbreaks that were recorded during the study period (black arrows). These landslide regions included 1) the Campos do Jordão, 08 landslides (beginning on 01/11/2013); 2) the São José dos Campos, 02 landslides (beginning on 01/10/2013) and 3) the Ubatuba, 03 landslides (beginning on 01/04/2013).

The climatic and geotechnical data that were recorded by the monitoring systems of the experimental field of Campos do Jordão (Ef-1) are shown in Figure 12. Here, the initial day (01/11/2013) of the 08 landslides recorded in the county during operation of the PPDC are indicated. By considering the rainfall data for the entire study period (Figure 12), the greatest rainfall concentration occurred between 12/01/12 and 04/30/13 (totaling 1832 mm of precipitation). This rainfall accounted for approximately 80% of the total rainfall that was recorded throughout the entire monitoring period (2334 mm), with a maximum rainfall intensity of approximately 30.0 mm/10 minutes on 02/15/13. When considering the early landslides on 01/11/13, the monitoring system of Campos do Jordão (Ef-1) recorded the following rainfall data (Figure 12, in detail): accumulated precipitation rates of 51.8 mm/24 h, 52.0 mm/48 h and 62.4 mm/72 h and a maximum hourly rainfall rate of 14 mm/h. By only analyzing the rainfall data (Figure 12), the accumulated rainfall of 51.8 mm/24 h was more significant. Thus, this rainfall accumulation was more effective for increasing the soil moisture content, which resulted in an outbreak of landslide events in Campos do Jordão. In addition, when analyzing soil moisture and temperature data (shown in Figure 12), the soil temperature values were among the highest that were recorded during the monitoring period on the day of landslide onset (black arrows) (20.6°C/0.3 m, 19.1°C/1.0 m, 17.7°C/2.0 m). In addition, the values of volumetric soil moisture were the highest among the values that were recorded throughout the study period (0.41 $m^3 \cdot m^{-3}$/0.3 m; 0.44 $m^3 \cdot m^{-3}$/0.6 m; 0.48 $m^3 \cdot m^{-3}$/1.0 m; 0.46 $m^3 \cdot m^{-3}$/1.5 m and 0.46 $m^3 \cdot m^{-3}$/2.0 m). By carefully analyzing the behavior of the moisture profile during the landslide periods (Figure 12), the soil moisture distribution throughout the soil profile was considerably different relative to other rainfall events. This difference was especially true when considering the drainage of the soil profiles (i.e., moments before the start of the landslides, all sensors recorded an increase in soil moisture that considered the delay measurement between the sensors due to their positions in the soil profile) before they reached their maximum values. These soils remained in this condition for approximately 2 h/0.3 m, 5 h/0.6 m, 2 h/1.0 m, 12 d/1.5 m and 30 h/2.0 m.

This spatial-temporal behavior distribution of moisture in the soil profile enhanced the soil temperature values and played an essential rule in the occurrences of landslides in the study area. This finding was demonstrated by the monitoring data of the other experimental fields (Figure 13and Figure 14), where 02 landslides were recorded in the municipality of São José dos Campos (starting day, 01/10/2013) and 03 landslides were recorded in Ubatuba (starting day, 01/04/2013).

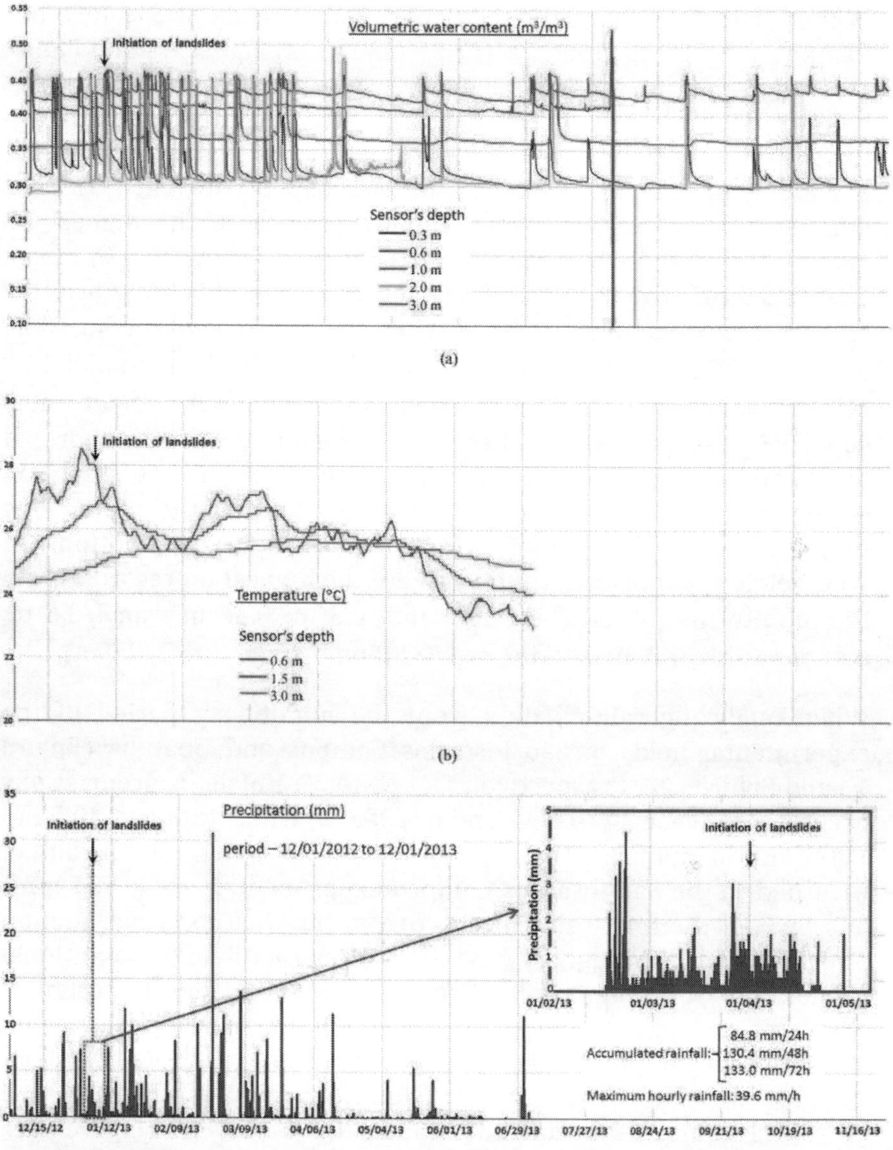

Figure 14. Monitoring data from the precipitation, soil temperature and soil moisture sensors in the experimental plot of Ubatuba (Ef-3) from 12/01/2012 to 12/01/2013, the main period of observed landslides (01/03/2013 to 01/04/2013, indicated by black arrows).

When analyzing data from the experimental fields of São José dos Campos (Ef-2) and Ubatuba (Ef-3) (shown in Figure 13 and Figure 14, respectively), similar behavior was observed in the spatial-temporal distribution of the soil moisture profile and high soil temperature values on days of landslide onset (black arrows). At the beginning of the landslides in the municipality of São José dos Campos (Figure 13), the soil temperatures were the highest on record during the monitoring period (26.1°C/0.6 m; 25.5°C/1.5 m; 23.4°C/3.0 m). The volumetric soil moisture contents were also the highest recorded during this period (0.43 $m^3 \cdot m^{-3}$/0.3 m; 0.39 $m^3 \cdot m^{-3}$/0.6 m; 0.44 $m^3 \cdot m^{-3}$/1.0 m; 0.34 $m^3 \cdot m^{-3}$/2.0 m and 0.28 $m^3 \cdot m^{-3}$/3.0 m), indicating changes in the moisture of the most superficial layers of soil (0.3 m, 0.6 m and 1.0 m). Furthermore, at the beginning of the landslides in Ubatuba (Figure 14), the soil temperatures were the highest on record during the monitoring period (28.0°C/0.6 m; 27.0°C/1.5 m; 25.0°C/3.0 m). In addition, the volumetric soil moisture contents were the highest on record during the monitoring period (0.45 $m^3 \cdot m^{-3}$/0.3 m; 0.46 $m^3 \cdot m^{-3}$/0.6 m; 0.46 $m^3 \cdot m^{-3}$/1.0 m; 0.45 $m^3 \cdot m^{-3}$/2.0 m; 0.46 $m^3 \cdot m^{-3}$/3.0 m).

When analyzing rainfall data from the monitoring period of the experimental fields of São José dos Campos and Ubatuba (Figure 13 and Figure 14, respectively), greater rainfall concentrations occurred between 12/01/12 and 04/30/13, which totaled 1207 and 1806 mm of rainfall, respectively. That is, 76% of the total rainfall recorded in the São José dos Campos experimental plot (1584 mm) and 89% of the total rainfall recorded in the Ubatuba experimental plot (1584 mm through 07/03/2013, after this date, no additional rainfall and soil temperature data were collected due to technical problems with the data logger). The maximum rain intensities that were recorded in the experimental fields of São José dos Campos and Ubatuba were approximately 28.0 mm/10 minutes on 12/30/12 (Figure 13) and 30.8 mm/10 minutes on 02/22/13 (Figure 14), respectively.

At the beginning of the landslides on 1/10/13, the São José dos Campos monitoring system (Ef-2) recorded the following rainfall data (Figure 13, in detail): accumulated precipitation of 89.4 mm/24 h, 90.6 mm/48 h and 90.8 mm/72 h and a maximum hourly rainfall of 56 mm/h. Based on the rainfall data (Figure 13), a rain accumulation of 89.4 mm/24 h with an intensity of 56 mm/h was potentially correlated to the outbreak of landslides in São José dos Campos. Regarding the monitoring system of Ubatuba (Ef-3), the landslide that occurred on 01/04/13 on highway BR 101 corresponded with the following rainfall data (Figure 14, in detail): accumulated precipitation of 84.8 mm/24 h, 130.4 mm/48 h and 133.0 mm/72 h and a maximum hourly rainfall of 39.6 mm/h. The monitoring data from the experimental field of Ubatuba (Figure 14) should be analyzed with caveats, mainly due to the considerable distance between the experimental and field locations of the landslides. However, based on the data that were recorded by the automatic environmental station installed next to the landslide location (Figure 10), very similar soil temperature and soil moisture values were recorded in the experimental plot of Ubatuba (Ef-3). Specifically, the average soil temperature and volumetric moisture contents were 28.5 to 28.0°C and 0.45 to 0.46 $m^3 \cdot m^{-3}$, respectively. For the values of accumulated precipitation in 24 h, a significant difference of approximately 40.0 mm (125.0 mm and 84.8 mm, respectively) occurred. The difference between the rain accumulation/24 h values that were recorded during the landslides in the monitored municipalities may indicate the importance of human activities in creating landslides.

Measured anomalies were temporarily or permanently displayed by some moisture sensors (10HS) in the experimental plots of Campos do Jordão (Figure 12, sensors installed at 1.0 and 1.5 m), São José dos Campos (Figure 13, sensor installed at 2.0 m) and Ubatuba (Figure 14, sensors installed at 1.0 and 2.0 m). These anomalies were due to the lack of temporary or permanent contact between the moisture sensors and the adjacent soil (predominantly clayey soil existing at these depths). The performance of volumetric moisture sensors (using the FDR technology—Frequency Domain Reflectometry) depends on the integrity of the moisture sensor's contact with the surrounding soil. Thus, the spaces (voids) that formed after the sensor installation in clayey layers (due to the shrinkage of the clay soil during the dry season) impaired the performance of these sensors for measuring the volumetric soil moisture content.

4. FINAL CONSIDERATIONS

The rupture mechanisms of landslides (Figure 5), geological and geotechnical characteristics, geomorphological conditions, existence of a deep water table, presence of unsaturated soil profile and the in situ monitoring data from the experimental plots (Ef-1, and Ef-2 and Ef-3) were considered. The environmental conditions that triggered landslides in the study areas were concurrent and included a significant increase in volumetric soil water content throughout the soil profile (up to 3.0 m deep), heavy rainfall (>35 mm/h) and/or cumulative rainfall of more than 50 mm/24 h and the high soil temperature values before heavy rains.

Based on the data analysis of the recorded soil moisture values in the experimental fields (Ef-1, Ef-2 and Ef-3), the critical values of soil moisture along the soil profiles where landslides were triggered were identified. For the studied soil profiles, the critical values of volumetric moisture were between 0.40 and 0.45 m^3/m^3, depending on the depth of the installed sensor. Furthermore, for landslides to occur, the volumetric moisture content must significant increase in the surface layers that are prone to soil rupture (usually between 1.0 to 2.0 m). In addition, the reading delay of the moisture sensors must be considered.

The accumulated rainfall data that were recorded in the municipalities of Campos do Jordão (Figure 12), São José dos Campos (Figure 13) and Ubatuba (Figure 10 and Figure 14) were analyzed. An inverse relationship was found between urban densification in risk areas (Figure 3) or the level of human intervention on natural slopes (for example, highway construction, Figure 9) and an increase in the amount of rainfall required to trigger landslides in the study area. For example, higher levels of human intervention on natural slopes (slope cuts/ landfills, release of wastewater, overload, etc.) reduce the amount of accumulated precipitation required for a landslide outbreak. The real-time monitoring of rainfall intensity (mm/h) and precipitation accumulation (mm/24 h) is essential for predicting landslides in urban or other risk areas, particularly when they are densely occupied.

Based on the temperature data of the soil profile (Figures 12-14), a significant relationship existed between the high soil temperature values (especially the superficial layers; H1/H2, Figure 4) and the landslide outbreaks that were recorded in the studied cities. This relationship may occur because the hydraulic behavior variations of the soil are directly proportional to the soil temperature variations (i.e., the permeability increases with increasing soil temperature [56]).

Regarding the soil suction data in the experimental plots (Ef-1, Ef-2 and Ef-3), no significant variations occurred during the monitoring period, mainly due to the unsatisfactory precision of the sensors (MPS-2 dielectric potential) for suction readings in the range of −10 to −100 kPa (error ±25%). However, the real-time monitoring of soil suction with sensors that are more accurate will be fundamental for understanding the mechanisms that trigger landslides.

The real-time and simultaneous monitoring of precipitation, soil moisture and temperature are essential for understanding the outbreak of landslides in risk areas, especially for allowing in situ and real-time assessment of spatial and temporal rainfall, soil moisture and temperature distributions. In addition, these data promise to establish critical thresholds of climatic and geotechnical variables for use in landslide warning systems in densely populated areas.

ACKNOWLEDGEMENTS

Rodolfo Moreda Mendes, Mário Valério Filho (process number 2011/22577-2).

REFERENCES

1. Zuquette, L.V. and Pejon, O.J. (2004) Geological and Geotechnical Hazardous Events in Brazil. In: Brazilian Symposium on Natural Disaster, Florianópolis-SC, GEDN/UFSC, 312-336.

2. Tatizana, C., Ogura, A.T., Cerri, L.E.S. and Rocha, M.C.M. (1987) Analysis of the Correlation between Rainfall and

Landslides in the Serra do Mar, Municipality of Cubatão. Proceedings of the Brazilian Congress of Engineering Geology, 2, 225-236.

3. Tatizana, C., Ogura, A.T., Cerri, L.E.S. and Rocha, M.C.M. (1987) Numerical Modeling of the Analysis of Correlation between Rainfall and Landslides Applied to the Slopes of the Serra do Mar in Municipality of Cubatão. Proceedings of the Brazilian Congress of Engineering Geology, 2, 237-248.

4. Lumb, P. (1975) Slope Failures in Hong Kong. Quarterly Journal of Engineering Geology and Hydrogeology, 8, 31-65.

5. Boonsinsuk, P., Yong, R.N. and Siu, S.K.H. (1982) Analysis of Hong Kong Residual Soil Slopes, Engineering and Constructions in Tropical and Residual Soil. ASCE Geotechnical Engineering Division, Honolulu, 463-482.

6. Zaruba, Q. and Mencl, V. (1982) Landslides and Their Control: Development in Geotechnical Engineering. Elsevier, Amsterdam, 324 p.

7. Brand, E.W., Premchitt, J. and Phillipson, H.B. (1984) Relationship between Rainfall and Landslides in Hong Kong. Proceedings of the 4th International Symposium on Landslides, Toronto, 16-21 September 1984, 377-384.

8. Anderson, M.G. and Pope, R.G. (1984) The Incorporation of Soil Water Models into Geotechnical Studies of Land- slides Behavior. Proceedings of the 4th International Symposium on Landslides, Toronto, 16-21 September 1984, 349- 353.

9. Canuti, P., Focardi, P. and Garzonio, C.A. (1985) Correlation between Rainfall and Landslides. Bulletin of the International Association of Engineering Geology, 32, 49-54.

10. Keefer, D.K., Wilson, R.C., Mark, R.K., Brabb, E.E., Brown, W.M., Ellen, S.D., Harp, E.L., Wieczorek, G.F., Alger, C.S. and Zatkin, R.S. (1987) Real-Time Landslide Warning during Heavy Rainfall. Science, 238, 921-925.

11. Eyles, N. and Howard, K.W.F. (1988) A Hydrochemical Study of Urban Landslides Caused by Heavy Rain: Scarborough

Bluffs, Ontario, Canada. Canadian Geotechnical Journal, 25, 455-466.

12. Matsuo, M. and Suzuki, H. (1988) Case Study of Slope Stability during Rainfall. Journal of Natural Disaster Science, 10, 15.

13. Wieczorek, G.F. (1996) Landslide Triggering Mechanisms. In: Turner, A.K. and Schuster, R.L., Eds., Landslides: Investigation and Mitigation, Transportation Research Board, National Research Council, Special Report, Washington DC, 76-90.

14. Anagnostopoulos, C. and Georgiadis, M. (1997) Analysis of Rainfall Data and Correlation to Landslides: The Case of Sykia-Pieria, Greece. In: Marinos, P.G., Koukis, G.C., Tsiambaos, G.C. and Stoumaras, G.C., Eds., Engineering Geology and the Environment, Balkema, Rotterdam, 483-487.

15. Au, S.W.C. (1998) Rain-Induced Slope Instability in Hong Kong. Engineering Geology, 51, 1-36.

16. Glade, T., Crozier, M.J. and Smith, P. (2000) Applying Probability Determination to Refine Landslide-Triggering Rainfall Thresholds Using an Empirical "Antecedent Daily Rainfall Model". Pure and Applied Geophysics, 157, 1059-1079.

17. Bacchini, M. and Zannoni, A. (2003) Relations between Rainfall and Triggering of Debris-Flow: A Case Study of Cancia (Dolomites, Northeastern Italy). Natural Hazards and Earth System Sciences, 3, 71-79.

18. Aleotti, P. (2004) A Warning System for Rainfall-Induced Shallow Failures. Engineering Geology, 73, 247-265.

19. Quinta Ferreira, M., Lemos, L.J.L. and Pereira, L.F.M. (2005) Influence of Rainfall on the Occurrence of Landslides in Coimbra, in the Last 139 Years. Portuguese Geotechnical Journal, 104, 17-30.

20. Chen, C.-Y., Chen, T.-C., Yu, F.-C., Yu, W.-H. and Tseng, C.-C. (2005) Rainfall Duration and Debris-Flow Initiated Studies

for Real-Time Monitoring. Environmental Geology, 47, 715-724.

21. Guidicini, G. and Iwasa, O.Y. (1976) Test of Correlation between Rainfall and Landslides in a Humid Tropical Environment. Technological Research Institute/IPT, Sao Paulo, 48 p.

22. Wolle, C.M. and Carvalho, C.S. (1989) Landslides on Slopes of Serra do Mar (Brazil). Soils and Rocks, ABMS/SPG: São Paulo, 12 (único): 27-36.

23. Wolle, C.M. and Hachich, W. (1989) Rain-Induced Land-Slides in Southern Brazil. Proceedings of the 12th International Conference on Soil Mechanics and Foundation Engineering, Rio de Janeiro.

24. Elbachá, A.T., Campos, L.E.P. and Bahia, R.F.C. (1992) Attempted of Correlation between Rainfall and Landslides in the City of Salvador. Proceedings of the Brazilian Conference on Stability of Slopes, 2, 647-673.

25. Almeida, M.C.J., Nakazawa, A. and Tatizana, C. (1993) Analysis of Correlation between Rainfall and Landslides in the City of Petrópolis, RJ. Proceedings of the Brazilian Congress of Engineering Geology, 2, 129-133.

26. Anjos, C.A.M., Cerri, L.E.S. and Gandolfi, N. (1997) Risk Situations and Measures to Prevent Accidents on Occupied Slopes in Maceio (AL), Brazil. Proceedings of the 2nd Pan-American Symposium of Landslides and Brazilian Conference on Slope Stability, Rio de Janeiro, 10-14 November 1997, 773-782.

27. Gusmão Filho, J.A. (1997) Urban Slopes: Environmental, Social and Political Aspects. Proceedings of the 2nd Pan-American Symposium of Landslides and Brazilian Conference on Slope Stability, Rio de Janeiro, 10-14 November 1997, 315-331.

28. Feijó, R.L., Paes, N.M. and d'Orsi, R.N. (2001) Rainfall and Mass Movements in the City of Rio de Janeiro. Proceedings of

the 3rd Brazilian Conference on Stability of Slopes, 1, 223-230.

29. Pinheiro, A., Tachini, M. and Riekmann, C.G. (2005) Study of the Correlation between Rainfall and Landslides in Risk Areas of Blumenau—SC. In: Vieira, R. and Pinheiro, A., Eds., A Look at the Risk Areas to Landslides in the City of Blumenau: Seeking Prevention, FURB-IPA, Blumenau, 27-41.

30. Macedo, E.S., Ogura, A.T. and Santoro, J. (2006) What Is a Contingency Plan or Preventive Plan of Civil Defense. In: Carvalho, C.S. and Galvão, T., Eds., Risk Prevention of Landslide in Slopes: Guide to Drafting of Municipal Policies, Ministério das Cidades, Brasília, 78-91.

31. Parizzi, M.G., Sebastião, C.S., Viana, C.S., Pflueger, M.C., Campos, L.C., Cajazeiro, J.M.D., Tomich, R.S., Guimarães, R.N., Abreu, M.L., Sobreira, F.G. and Reis, R. (2010) Correlations between Rainfall and Mass Movements in the City of Belo Horizonte, MG. Geografias, 6, 49-68.

32. Fisch, G., Vendrame, I.F. and Hanaoka, P.C.M. (2007) Rainfall Space Variability during the LBA/TRMM Experiment 1999 in Amazônia. Acta Amazonica, 37, 583-590.

33. Santoro, J., Mendes, R.M., Pressinotti, M.M.N. and Manoel, G.R. (2010) Correlation between Rainfall and Landslides Occurred during the Operation of the Prevention Plan of Civil Defense in State of São Paulo, SP. Proceedings of the 7th Brazilian Symposium on Geotechnical and Geoenvironmental Cartography, Maringá, 1-15.

34. Mendes, R.M., Marinho, F.A.M. and Orlando, P.G. (2006) Geotechnical Properties of Unsaturated Residual Soils of the Serra do Mar. Proceedings of the 13th Brazilian Congress of Soil Mechanics and Geotechnical Engineering, Curitiba.

35. Silva, P.C.F., Mendes, R.M., Nery, T.D. and Silva, I.C. (2007) Spatial Analysis of Geotechnical Properties of the Soils of Ubatuba by Local Deterministic Modeling and Geostatistics. Proceedings of the 6th Brazilian Symposium on Geotechnical

and Geoenvironmental Cartography, Uberlândia-MG, ABGE, 309-323.

36. Mendes, R.M., Marinho, F.A.M. and Prado, R.L. (2008) Use of the GPR Method for the Monitoring of the Moisture on Slopes. Proceedings of the 14th Brazilian Congress of Soil Mechanics and Geotechnical Engineering, Búzios.

37. Mendes, R.M. and Marinho, F.A.M. (2008) Evaluation of a Residual Soil Profile of the Serra do Mar by Porosimetry. Proceedings of the 12th Brazilian Congress of Engineering Geology and Environmental, Porto de Galinhas.

38. Mendes, R.M. (2008) Study of the Geotechnical Properties of Unsaturated Residual Soils of Ubatuba (SP). Doctoral Thesis, Polytechnic School, Sao Paulo University, Sao Paulo, 255 p.

39. Lima Junior, S.B., Prado, R.L. and Mendes, R.M. (2006) The Use of the Rayleigh Waves for in Situ Estimates of the Wave Velocities of Unsaturated Soils in Areas Subject to Mass Movements. Proceedings of the 2nd Brazilian Symposium on Geophysics, Natal, 1-10.

40. Prado, R.L., Mendes, R.M. and Marinho, F.A.M. (2007) Estimates of Soil Moisture Content Using the GPR Method on Landslide Area in Ubatuba/SP. Proceedings of the 10th International Congress of the Brazilian Geophysical Society, Rio de Janeiro, 19-22 November 2007.

41. Mendes, R.M., Marinho, F.A.M. and Funari, A.H. (2007) Evaluation of Granular Matrix Sensor (GMS) for Measuring the Suction. Proceedings of the 6th Brazilian Symposium of Unsaturated Soil, Salvador, 1-3 November 2007.

42. Barbosa, E.E.M., Prado, R.L., Marinho, F.A.M., Oliveira, O.M. and Mendes, R.M. (2008) Evaluation of GPR Method to Estimate the Moisture Content of Sediments in Controlled Test. Proceedings of the 3rd Brazilian Symposium on Geophysics, Belém, 1-5.

43. Barbosa, E.E.M., Prado, R.L., Mendes, R.M. and Marinho, F.A.M. (2010) Estimation of Moisture Content Using the GPR

Method: A Comparative Evaluation in Laboratory and Field Experiments. Brazilian Journal of Geophysics, 28, 691-701.

44. Vargas, M. (1999) Historical and Conceptual Review of the Landslides of the Serra do Mar. Soil and Rocks, 22, 53-83. (In Portuguese) [Paper reference:1]

45. DAEE/Department of Water and Electric Power of the State of Sao Paulo (2013) Rainfall Database of the State of Sao Paulo (Brazil). Open Access Publisher.

46. Modenesi-Gauttieri, M.C., Toledo, M.C.M., Hiruma, S.T., Taioli, F. and Shimada, H. (2011) Deep Weathering and Landscape Evolution in a Tropical Plateau. Catena (Cremlingen), 85, 221-230.

47. Ahrendt, A. and Zuquette, L.V. (2003) Triggering Factors of Landslides in Campos do Jordão City, Brazil. Bulletin of Engineering Geology and the Environment, 62, 231-244.

48. Ahrendt, A. (2005) Gravitational Mass Movements—Proposal of a Forecast System: Application at the Urban Area of Campos do Jordão City-SP—Brazil. Doctoral Thesis, School of Engineering of São Carlos, São Paulo University, São Carlos, 390 p.

49. Heilbron, M., Pedrosa-Soares, A.C., Neto, M.C.C., Silva, L.C., Trouw, R.A.J. and Janasi, V.A. (2004) Mantiqueira Province. In: Geology of the Sul-American Continent: Evolution and Work of Fernando Flávio Marques de Almeida, 203-236.

50. Hasui, Y., Mioto, J.A. and Morales, N. (1994) Geology of the Pré-Cambriano. In: Falconi, F.F. and Negro Jr., A., Eds., Soils of the Coast of Sao Paulo, ABMS, Sao Paulo, 41-68.

51. Morais, S.M., Rodrigues, J.B., Sachsi, L.L.B. and Dehler, N.M. (1999) Integrating the Geological Map of Santos 1:250.000, State of Sao Paulo (Brazil). Ministério de Minas e Energia—CPRM, Serviço Geológico do Brasil.

52. Wolle, C.M. (1985) Peculiarities of "in Situ" Behavior of Tropical Soils: Slope Stability. Committee on Tropical Soils. Proceedings of the 1st International Conference on

Geomechanics in Tropical Lateritic and Saprolitic Soils, Brasília, Progress Report. t.3.2, 51 p.

53. Wolle, C.M., Guidicini, G., Araujo, J.S. and Pedrosa, A. (1978) A Slide Mechanism in the Slopes of the Serra do Mar Southern Brazil. Proceedings of the 3rd International Conference of Engineering Geology, Madrid, 4-8 September 1978, 14 p.

54. CEDEC/Coordination of Civil Defense of the State of Sao Paulo (2013) Preventive Plan of Civil Defense-PPDC: Database 2012/2013.

55. INPE/National Institute of Spatial Research (2014) Integrated System of Environmental Data/SINDA—Data of Automatic Environmental Stations—PESM Picinguaba/Ubatuba (SP).

56. Habibagahi, K. (1973) Temperature Effect and the Concept of Effective Void Ratio. Indian Geotechnical Journal, 7, 14-34.

8

CHAPTER

Nonlinear Seismic Vulnerability Evaluation of Irregular Steel Buildings with Cumulative Damage Indices

Mohsen Gerami, Yahya Sharbati* and Abbas Sivandi-Pour

Faculty of Civil Engineering, Semnan University, Semnan 3513119111, Iran

ABSTRACT

Measuring structural damage during earthquakes has always been a challenging problem for earthquake engineers. Various damage indices are proposed with the objective of quantifying the structural damage in prototype and model structures subjected to seismic excitation. In this study, seismic vulnerability of irregular steel buildings is assessed in three dimensions considering effects of the panel zone, which has not been considered in recent studies of the field of seismic vulnerability. The buildings are modeled with different storeys and irregular plans. Seismic performance of buildings was assessed in life-safety and collapse-prevention levels. Cumulative functions of damage indices are applied in the nonlinear dynamic analysis of buildings in the near-field ground motions. It is concluded that participation rates of deformation and energy in the damage of irregular buildings are 74.5% and 25.5%, respectively. Severe damage and collapse due to seismic dissipated energy occurred in the initial storeys of low-rise buildings and in the middle storeys of high-rise buildings.

Keywords: Seismic vulnerability Damage index Performance level Irregular buildings Panel zone

INTRODUCTION

Damage and detraction in recent earthquakes have shown that even though buildings designed according to recent codes and regulations have performed well from the viewpoint of the safety of human lives, the level of damage to the buildings and the consequent economic losses are unexpectedly high. In order to assess the reliability of structures subjected to ground motions, it is necessary to evaluate failure modes, which lead to cyclic deterioration in strength, stiffness, and energy dissipation. Many researchers have tried to determine numerical functions in order to express the partial and total vulnerability of structures. These functions are considered as the first step of the seismic rehabilitation of the structure. More recent damage functions showed the more realistic view to structural damage. One of the main topics of debate in vulnerability evaluation of structures is selecting the most appropriate damage functions for a special structure to show a structural situation more realistically (Foutch and Yun 2002; Kalkan and Kunnath 2007; Gerami and Sivandi-Pour 2008; Nethercot 2011). In a study, Elnashai (2006) evaluated seismic vulnerability of structures. He introduced different methods and tools for evaluating vulnerability of buildings. Shibin et al. (2010) assessed vulnerability and control of structures based on the performance levels. They suggested a probable approach for seismic evaluation of buildings. (Bojórquez et al. 2010) introduced an energy damage index for multi-degree-of-freedom steel buildings. Their model was developed by complementing the results obtained by experimental and analytical investigation on steel frame elements regarding the distribution of plastic demands on several steel frames. Pitocco (2011) studied the information technology in the context of evaluation of steel building vulnerability. He suggested some methods for modification of damaged buildings. Determination of most damage indices involves complicated and time-consuming computations that are neither economical nor feasible in concurrent structural engineering practice. Cumulative damage indices are usually modeled either by using a low-cycle fatigue formulation, in which damage is taken as a function of the accumulated plastic deformation, or by inserting a term related to the dissipated hysteretic energy in the damage model (Kunnath and Chai 2004; Poljanšek and Fajfar 2008; Karimi et al. 2011). No study was found on seismic vulnerability of buildings considering panel zone and irregularity. In this study, the panel zone is considered and irregular buildings are modeled three dimensionally. Cumulative damage indices considering deformation and dissipated energy (ductility, energy, Park-Ang) were used. To evaluate the seismic vulnerability of models, near-field records of Tabas, Northridge, and Kobe earthquakes were used.

METHOD

Design of buildings

Models in this study include steel buildings with three bays in five types of 4, 7, 10, 15, and 20 storeys. The height of each storey was 3.5 m, and the length of each bay is 4.5 m. The buildings were designed based on the Iranian code of practice for seismic-resistant design of buildings (Standard 2800) and LRFD method of AISC (2005) (see Appendix). The loads and load combinations were determined according to ASCE/SEI 7-05 (2005). The buildings have irregular plans as shown in Figure 1.

A lateral resistant system is a moment frame in the *X*-direction and a dual system of moment frame and bracing in the *Y*-direction. IPE sections for beams, IPB sections for columns, and UNP sections for braces were used in the modeling and design.

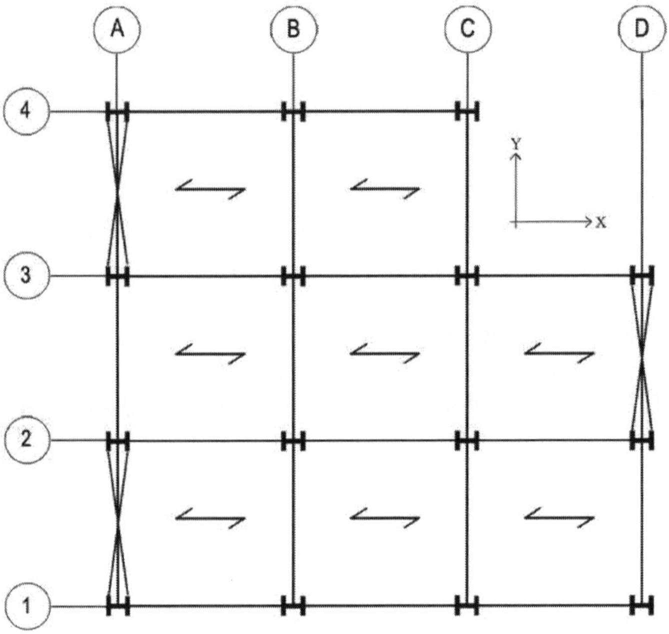

Figure 1 Plan of buildings.

Nonlinear modeling of elements

For nonlinear dynamic analysis, using OpenSees, an open-source finite element platform has been developed at the University of California-Berkeley for earthquake performance assessment (Mazzoni et al. 2006). Fy, *E*, *G*, and strain hardness were considered to be 2,400 kg/cm^2, 2.1 × 10^6, 807,692.30, and 3%, respectively. For nonlinear dynamic analysis, a nonlinear beam column element is used, which is based on the non-iterative (or iterative) force formulation, and considers the distribution of plasticity along the element. In order to do nonlinear dynamic analysis, each element was divided into ten parts. In these parts, the stresses and strains were derived from three points at the top, middle, and bottom corners of the section by which cyclic energy dissipation in each frame element was calculated.

Modeling of connections and panel zone

During seismic events, connections in steel buildings would experience a large amount of stress and deformation demands due to their critical position in the structure. Panel zone is the web area of beam-to-column connection delineated by the extension of beam and column flanges through the connection, transmitting moment through a shear panel. Panel zones are assigned to joints to model the flexibility of beam-column connections. Correct modeling of the beam-column connections is very important in the nonlinear analysis of moment frames. Four models were used in the steel buildings in order to model the connections of beam to column:

1. *Linear model*. This model is suitable for designing a moment-resisting frame. Although the model shows reasonable results for design, it cannot accurately forecast the distribution of the inelastic frame element forces.

2. *Elastic model with modeling panel zone*. The beam and column will be connected to the panel zone by a rigid link. Stiffness of the panel zone is modeled as a spring.

3. *Nonlinear model*. Yielding of frame elements is higher than that of the elastic model. This method is currently used in the modeling of buildings. Modeled springs will remain rigid until the section reaches plastic moment.

4. *Nonlinear model with modeling panel zone*. Three approaches can be used in the modeling of the panel zone. The first is to join with a nonlinear spring. In the second one, two springs are used: one for rigidity of the panel zone and the other for average strength of beam and the panel zone. In the third approach, the panel zone is modeled with eight rigid elements (Krawinkler 2000; Lignos 2008; Kim and Engelhardt 2002; Asgarian et al. 2010).

In this study, a nonlinear model with modeling of the panel zone was used in modeling the joints in OpenSees. Two nonlinear springs were modeled. Beams and columns are jointed by the rigid link in the panel zone. Although the rigid link increases joint stiffness, the modeled spring in the panel zone causes flexibility. A rotational spring with trilinear behavior is then utilized to tie the beam and column together. The rigid links stiffen the structure, but the panel zone spring adds flexibility. The net result would lead to a relatively stiffer assembly than that of the center-line model. Since it is stiffer, it would help in satisfying the drift design criteria. This model has been shown in Figure 2.

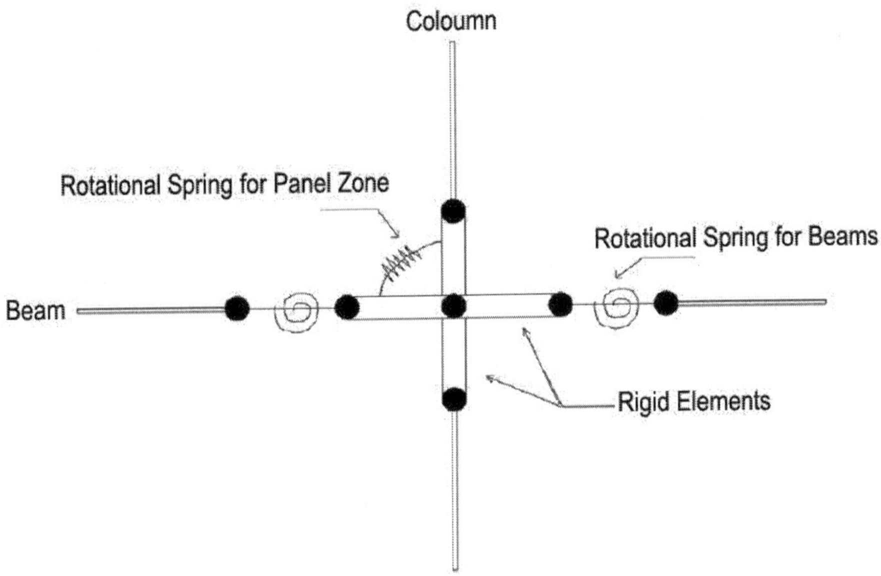

Figure 2 Modeling of panel zone in OpenSees.

The accurate description of the panel zone stiffness is very important in the analyses leading to the estimation of the fundamental period and of the expected frame drift. Spring stiffness of the panel zone is obtained from the following relation:

$$K_\theta = \frac{M_y}{\theta_y} \tag{1}$$

$$M_y = V_y.d_b = 0.55F_y d_c t d_b \tag{2}$$

$$\theta_y = \gamma_y = \frac{F_y}{\sqrt{3}G} \tag{3}$$

where F_y represents the yield strength of panel zone, G represents the shear module, $d_c$represents the column height, d_b is beam height, and t is thickness of panel zone.

Near-field ground motion records

Near-field problems have become an important topic for both seismologists and earthquake engineers. The characteristics of near-field records of an earthquake are quite different from the usual far-field records. After the original recognition of their differences in the Port Hueneme earthquake in 1957, a lot of inhabited structures and lifeline systems were damaged in the major earthquakes which happened in the following year. The special characteristics of near-field ground motions are directly related to the earthquake source mechanism, rupture direction relative to the site, and slip direction of the rupture fault (Gioncu 1998; Tirca et al. 2003; Alavi and Krawinkler 2004; Brun et al. 2004; Bray and Rodriguez-Marek 2004; Li and Xie 2007; Gerami et al. 2012; Monavari and Massumi 2012). In the present research work, three near-field earthquake records (Tabas, Kobe, and Northridge) were chosen. Models were loaded under longitudinal and transversal components of each record. Records were analyzed, and the difference between damages and function of different damage indices in each storey of the buildings were evaluated and compared. The specifications of records are presented in Table 1and Figure 3.

Table 1 Specification of ground motions

Earthquake	Location	Date	PGA (g)		Ms
			Long	Trans	
Tabas	Iran	1978	0.85	0.83	7.8
Kobe	Japan	1995	0.75	0.36	6.9
Northridge	USA	1979	0.82	0.6	6.5

PGA, peak ground acceleration; Long, longitudinal; Trans, transversal.

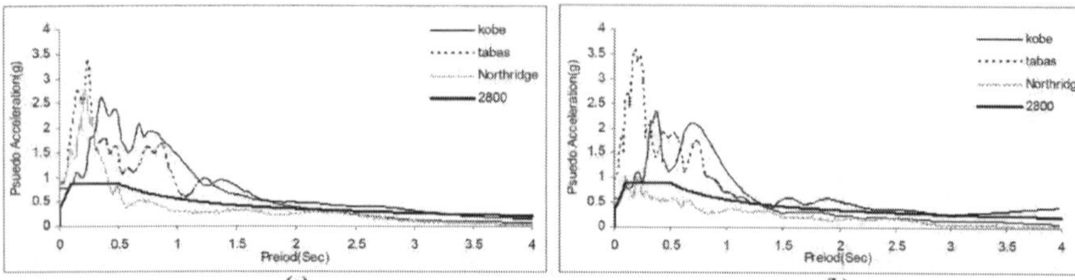

(a) (b)

Figure 3 Spectral accelerations of ground motion records. (a) Longitudinal component and (**b**) transversal component.

Seismic performance assessment

Two hazard levels of 1 and 2 were defined for seismic performance evaluation and structural vulnerability. The hazard level 1 (Basic Safety Earthquake (BSE)-1) is determined based on 10% probability of the event during 50 years that is equivalent to a return period of 475 years. The hazard level 2 (BSE-2) is determined based on 2% probability of the event during 50 years that is equivalent to a return period of 2475 years.

Relative displacement of the storeys is one of the general standards used in studying the performance of structures. The maximum relative displacement of a building's roof in each different level for different earthquakes has been presented in Table 2.

Table 2 Maximum relative displacement in two hazard levels

Models	Maximum displacement (cm)					
	BSE-1			BSE-2		
	Tabas	Northridge	Kobe	Tabas	Northridge	Kobe
4	19.8	20	20.5	21.5	29.5	22.7
7	17.48	29.72	26.14	27.78	38.2	35.05
10	44.6	36.91	51.05	73.32	49.91	80.16
15	59.51	44.9	62.68	88.82	45.33	78.56
20	63.35	44.51	56.69	117.94	58.19	64.46

If the frame elements are controlled by deformation, this index can be calculated. The beams are assumed to be displacement-controlled frame elements. As with the columns, if the ratio ofP/P_{cl} is less than 0.5, the control is measured by displacement. In the parameters required in the determination of ductility, yield and maximum rotation of frame elements were considered. Maximum rotation of frame elements in life safety (LS) and collapse prevention (CP) performance levels is given in Table 3.

Table 3 Maximum rotation of frame elements in performance levels

Models	Maximum rotation of columns (rad)				Maximum rotation of beams (rad)
	Uniform distribution		Spectral distribution		
	LS	CP	LS	CP	
4	0.008	0.014	0.005	0.008	0.0097
7	0.005	0.007	0.008	0.005	0.0097
10	0.0063	0.0090	0.0219	0.0106	0.011
15	0.0028	0.0034	0.0036	0.0046	0.0079
20	0.0031	0.0038	0.0049	0.0064	0.0079

The determination of target displacement of models using coefficient method according to FEMA 356 (2000) is shown in Table 4. This method estimates the target displacement, providing a direct numerical process for calculating the displacement demands.

Table 4 Target displacement of models in performance levels

Models	Target displacement in *X*-direction (cm)				Target displacement in *Y*-direction (cm)			
	Uniform load pattern		Dynamic load pattern		Uniform load pattern		Dynamic load pattern	
	LS	CP	LS	CP	LS	CP	LS	CP
4	17.87	32.92	20.10	32.92	5.80	9.49	5.80	9.49
7	33.03	54.04	39.91	65.30	13.33	21.82	13.33	21.82
10	37.88	61.98	40.58	66.41	22.16	36.26	22.16	36.26
15	48.05	78.63	58.07	95.02	44.02	72.04	44.02	72.04
20	57.51	94.11	69.49	113.71	59.18	96.85	59.18	96.85

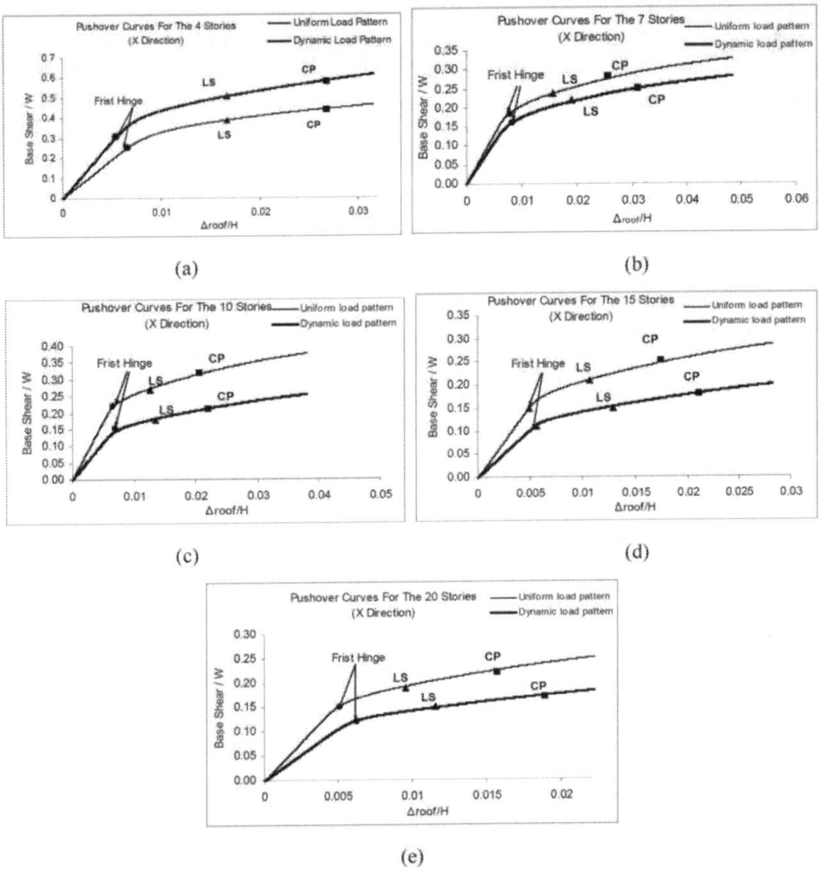

Figure 4 Pushover curve of models. (**a**) 4, (**b**) 7, (**c**) 10, (**d**) 15, and (**e**) 20 storeys.

Buildings were loaded with live and dead loads for gravity loading, and then laterally loaded. For each building, the total weight of the building, total height, fundamental translational mode period, roof displacement, and base shear were recorded. Figure 4 illustrates the capacity curve of models in the X-direction under uniform and dynamic load pattern.

In Figure 4, W represents the weight, H represents the height, and Δroof represents the maximum displacement of models. Maximum displacement of models in BSE-1 and BSE-2 are shown in Figure 5.

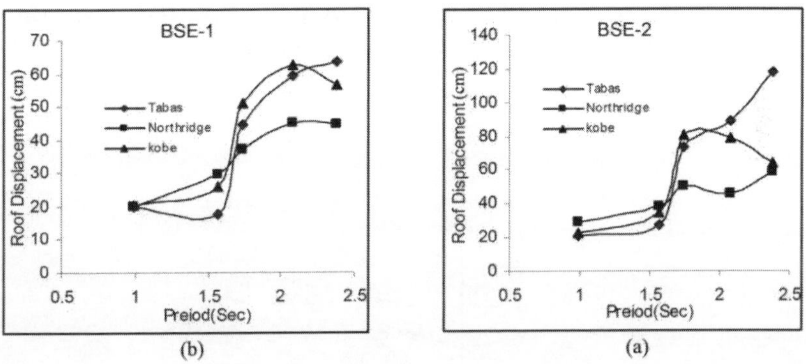

Figure 5 Maximum roof displacement variations in (a) BSE-1 and (b) BSE-2.

RESULTS

Damage index is known to be a standard measure in evaluating structural damage. Damage indices are usually normalized so that their value is equal to zero when there is no damage and is equal to unity when total collapse or failure occurs. Damage indices can involve a combination of one or more variables in its determination. Damage indices are applicable to different types of structural systems under different loadings and are defined based on parameters indicating economic conditions.

Ductility damage index

The simplest definition for a damage function belongs to ductility damage index. Energy dissipation by elements is not considered in ductility damage index, and only maximum deformation is used while expressing the damage amount. This parameter shows the ratio of maximum deformation and yield deformation of frame elements. Equation 4 shows the ductility damage index.

$$DI_{Ductility} = \frac{\theta_m}{\theta_y} \tag{4}$$

In this relation, θ_m represents the maximum rotation of frame elements during earthquake and θ_y denotes the yield rotation of elements. Figure 6 shows a schematic curve for calculating the ductility factor.

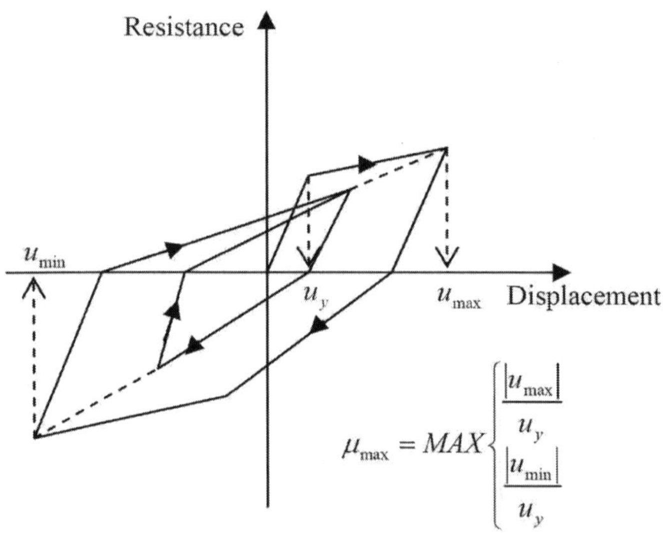

Figure 6 Maximum ductility factor.

A ductility factor of less than 1 indicates elastic response, and the values of more than 1 show non-elastic responses, if there is monotonic loading. This damage index is presented in Figures 7, 8, and 9.

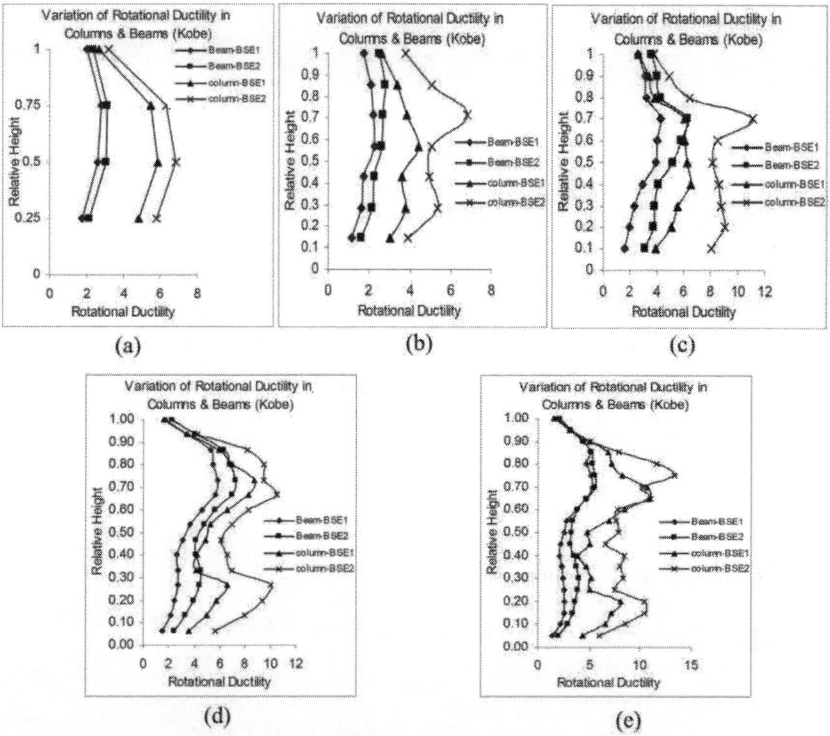

Figure 7 Ductility damage index in Kobe earthquake. For models (**a**) 4, (**b**) 7, (**c**) 10, (**d**) 15, and (**e**) 20 storeys.

Maximum value of the ductility damage index in frame elements is presented in Table 5. The mean value of ductility damage indices in beams in BSE-1 and BSE-2 are 4.3 and 6.4, respectively. This value is 8.3 for columns and 11.9 for BSE-1 and BSE-2. From the average values, it can be conclude that the deformation of the columns is 93% higher than the BSE-1 beams and 87% higher than BSE-2.

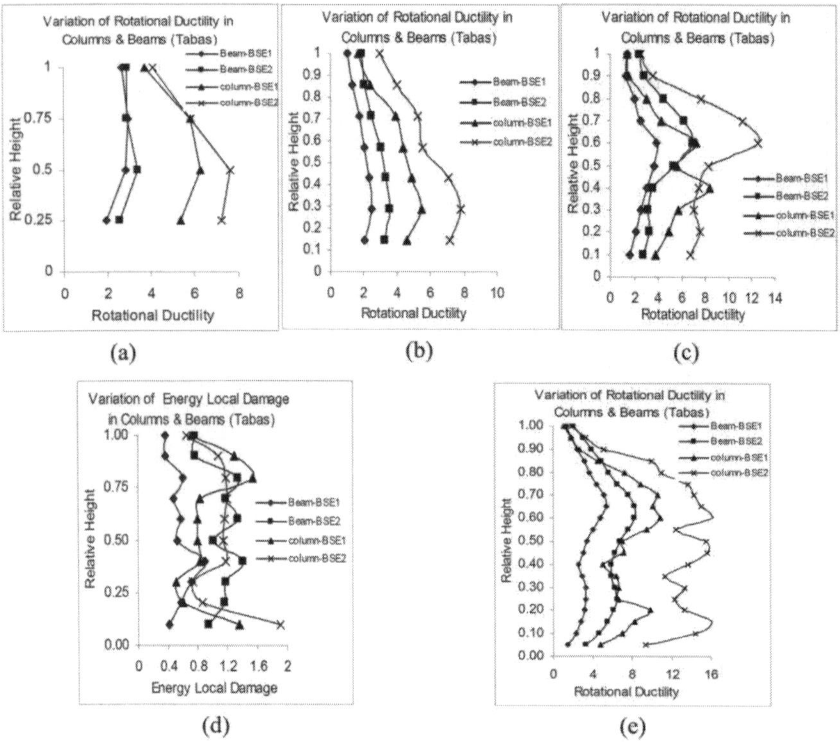

Figure 8 Ductility damage index in Tabas earthquake. For models (**a**) 4, (**b**) 7, (**c**) 10, (**d**) 15, and (**e**) 20 storeys.

Energy damage index

The amount of energy dissipation in each frame element is considered as an index in the evaluation of performance and structural damage. This function is defined based on hysteretic energy dissipated by elements. Energy damage index considers only the energy absorbed by elements and does not include their deformations. Regarding energy dissipated in damage index function, the effects of earthquake duration and damage aggregation are included. Energy damage index is expressed according to the following relation:

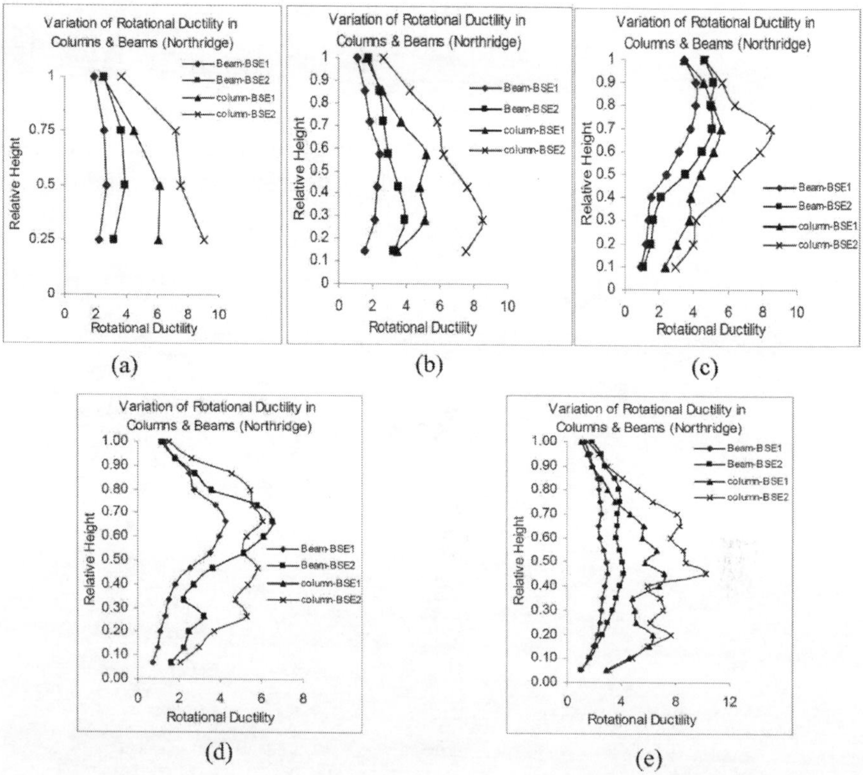

Figure 9 Ductility damage index in Northridge earthquake. For models (**a**) 4, (**b**) 7, (**c**) 10, (**d**) 15, and (**e**) 20 storeys.

Table 5 Maximum value of ductility damage index in frame elements

Models	Beams		Columns	
	BSE-1	BSE-2	BSE-1	BSE-2
4	2.92	3.91	6.3	9.02
7	2.51	3.89	6.12	8.55
10	4.32	6.98	8.39	12.55
15	6.43	8.97	9.48	13.39
20	5.39	8.29	10.88	16.08
Average	4.30	6.40	8.30	11.90

$$\mathrm{DI}_{\mathrm{Energy}} = \frac{\mathrm{EH}/F_y\Delta_y}{(\Delta_u/\Delta_y - 1)} \tag{5}$$

where EH is the energy dissipated by the system, and Δ_y is yield limit strength of the frame element. Energy damage indices of the models have been compared for different earthquakes in Figures 10, 11, and 12.

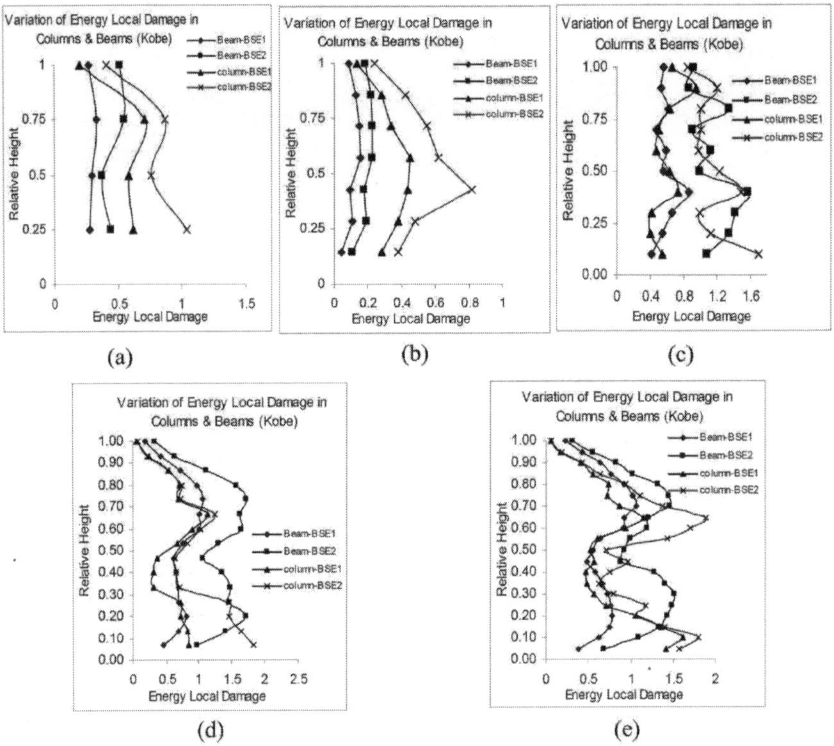

Figure 10 Energy damage index in Kobe earthquake. For models (**a**) 4, (**b**) 7, (**c**) 10, (**d**) 15, and (**e**) 20 storeys.

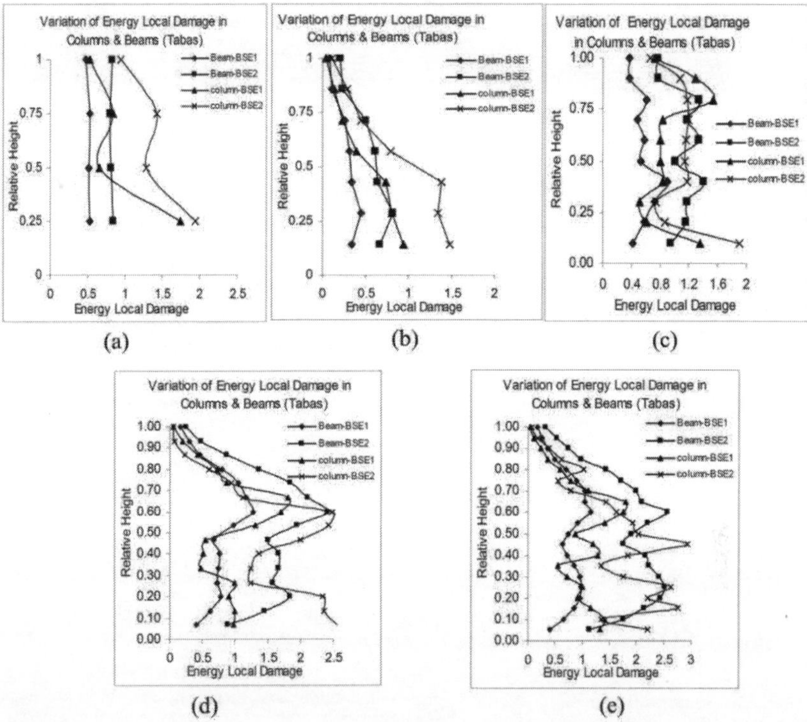

Figure 11 Energy damage index in Tabas earthquake. For models (**a**) 4, (**b**) 7, (**c**) 10, (**d**) 15, and (**e**) 20 storeys.

Results of this index showed that most damages in low-rise buildings occurred in initial storeys and in high-rise buildings occurred in intermediate storeys. The maximum value of energy damage index in frame elements is given in Table 6.

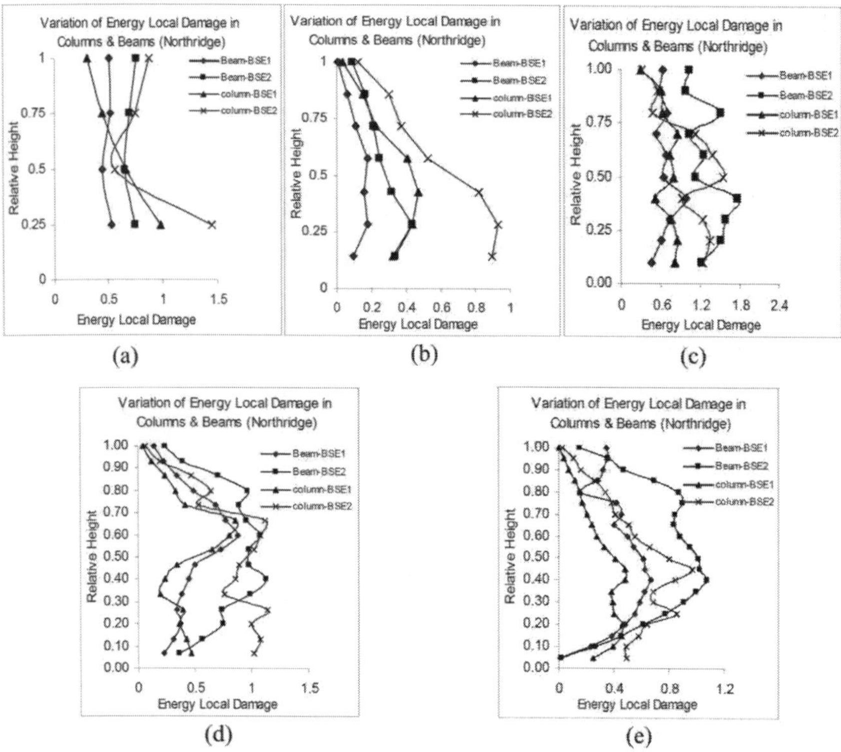

Figure 12 Energy damage index in Northridge earthquake. For models (**a**) 4, (**b**) 7, (**c**) 10, (**d**) 15, and (**e**) 20 storeys.

Table 6 Maximum value of energy damage index in frame elements

Models	Beams		Columns	
	BSE-1	BSE-2	BSE-1	BSE-2
4	0.53	0.84	0.7	0.86
7	0.45	0.35	0.82	0.94
10	0.97	0.99	1.53	1.89
15	1.28	2.41	1.81	2.55
20	1.15	2.56	1.18	2.93

The mean value of energy damage indices in beams in BSE-1 and BSE-2 are 0.87 and 1.43, respectively. For the columns, these values increase to 1.2 and 1.83 for BSE-1 and BSE-2. It can be observed that energy dissipated by columns in earthquakes is 38% higher in BSE-1 and 28% higher in BSE-2 compared to the beams.

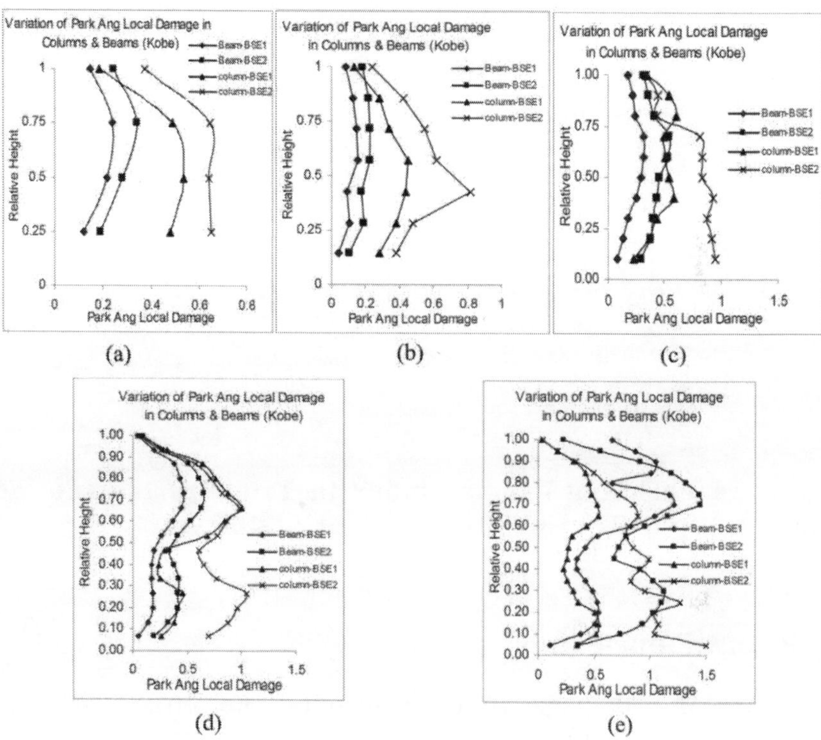

Figure 13 Park-Ang damage index in Kobe earthquake. For models (**a**) 4, (**b**) 7, (**c**) 10, (**d**) 15, and (**e**) 20 storeys.

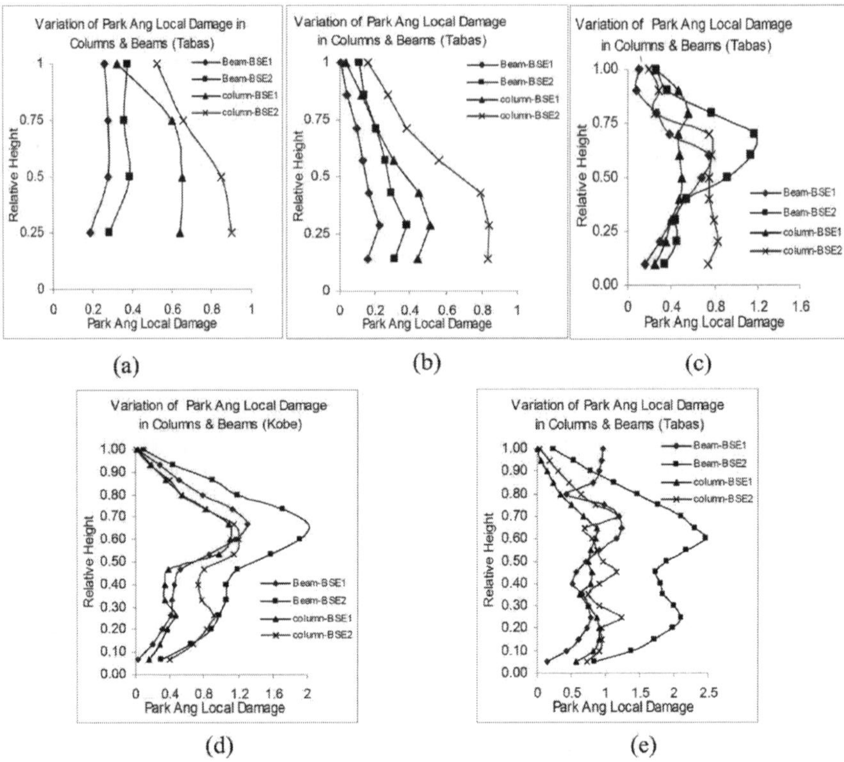

Figure 14 Park-Ang damage index in Tabas earthquake. For models (**a**) 4, (**b**) 7, (**c**) 10, (**d**) 15, and (**e**) 20 storeys.

Park-Ang damage index

The Park-Ang damage index considers the effects of both parameters of maximum deformation and dissipated energy in damage evaluation (Park et al. 1984).

The relation of this index can be expressed as follows:

$$\mathrm{DI}_{\mathrm{Park-Ange}} = \frac{\Delta_m}{\Delta_u} + \beta \frac{\mathrm{EH}}{F_y \Delta_u} \tag{6}$$

where Δ_m is the maximum deformation of the element, Δ_u is the ultimate deformation, and β is a model constant parameter.

In this research, the value of β is 0.15, indicating the amount of energy dissipated in the damage range. This damage index is one of the functions that offer a certain range for describing the physical concept of the damage level. A comparison of this damage index for different earthquakes is shown in Figures 13, 14, and 15.

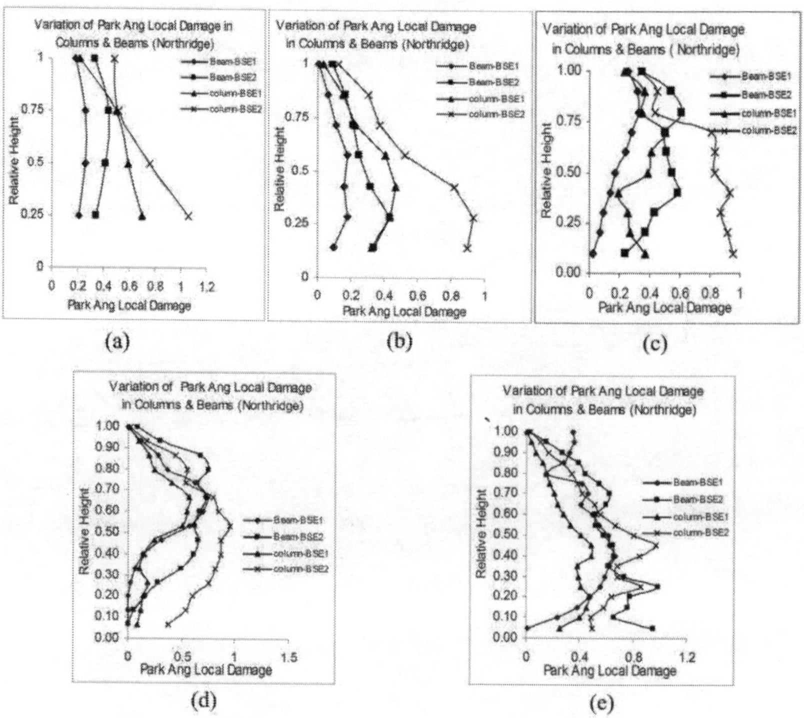

Figure 15 Park-Ang damage index in Northridge earthquake. For models (**a**) 4, (**b**) 7, (**c**) 10, (**d**) 15, and (**e**) 20 storeys.

Maximum value of the Park-Ang damage index in frame elements is presented in Table 7.

Table 7 Maximum value of Park-Ang damage index in frame elements

Models	Beams		Columns	
	BSE-1	BSE-2	BSE-1	BSE-2
4	0.28	0.43	0.65	0.9
7	0.22	0.43	0.47	0.95
10	0.75	1.18	0.61	0.96
15	1.3	1.79	0.11	1.21
20	1.24	2.47	0.91	1.49

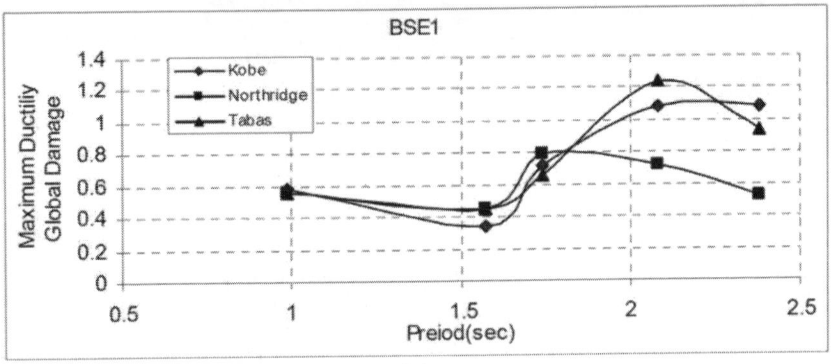

(a)

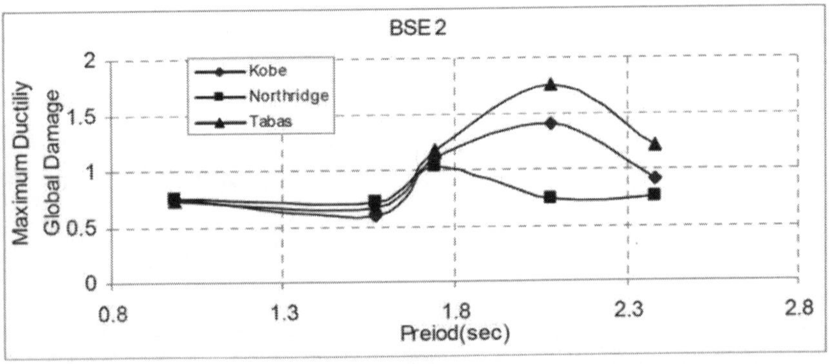

(b)

Figure 16 Maximum ductility damage index variations of storeys in (a) BSE-1 (b) BSE-2.

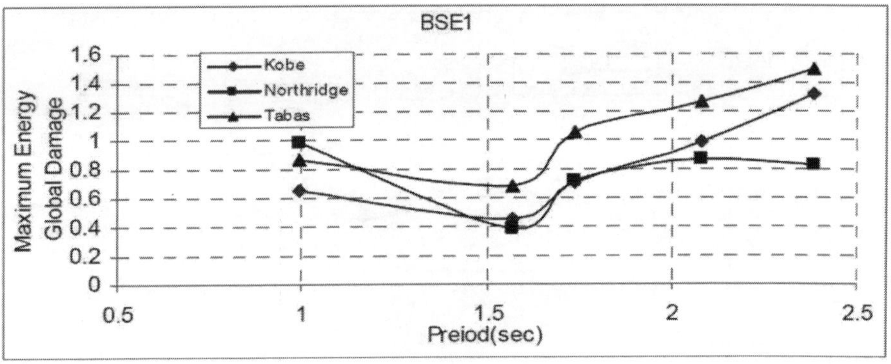

(a)

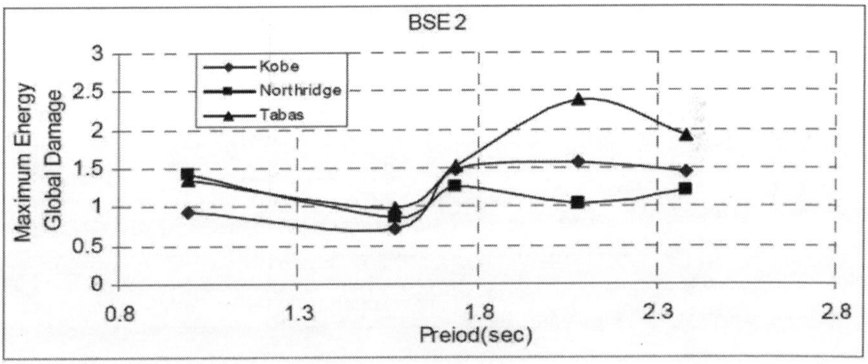

(b)

Figure 17 Maximum energy damage index variations of storeys in (a) BSE-1 (b) BSE-2.

The average of the Park-Ang damage index in beams for BSE-1 and BSE-2 is 0.75 and 1.26, respectively. This value is 0.55 and 1.1 for columns.

DISCUSSION

Maximum values of damage indices are shown in Figures 16, 17, and 18.

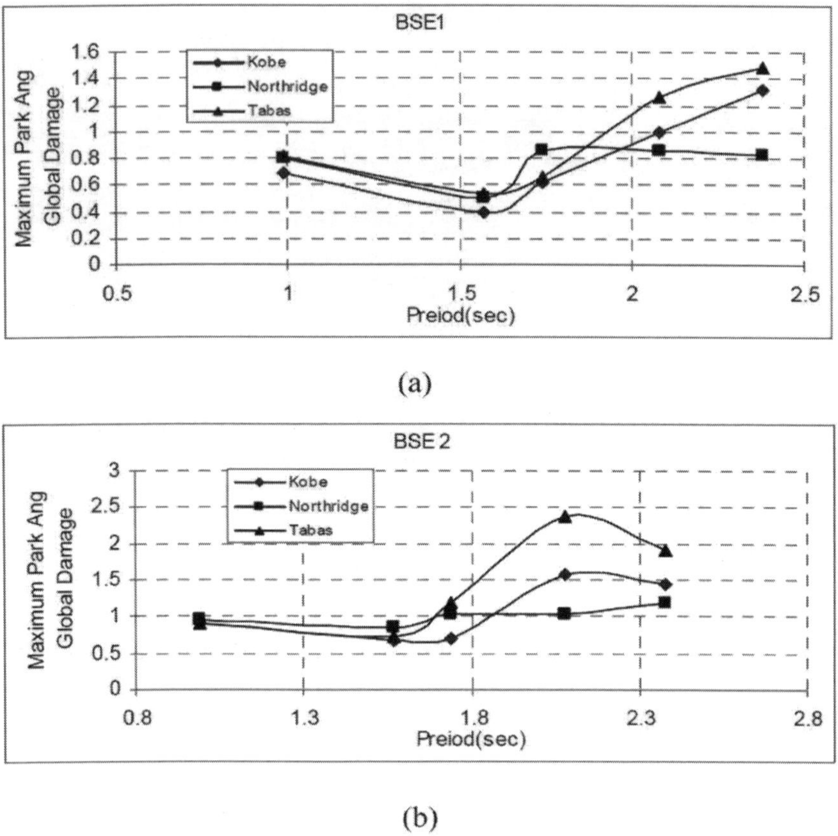

Figure 18 Maximum Park-Ang damage index variations of storeys in (a) BSE-1 (b) BSE-2.

There was no remarkable difference in response of buildings under records for 4- and 7-storey buildings. Increase in storeys caused noticeable changes in the results of records, so this difference will reach a peak in the 15-storey model. The maximum value of ductility damage index belonged to the 15-storey model in hazard level 1, and its value was 1.23 for Tabas earthquake. In hazard level 2, the highest value of this index was equal to 1.74, similar to the first hazard level.

As with the energy damage index, the highest energy absorption belonged to the 20-storey model under Tabas earthquake in hazard level 1. The minimum value of energy absorption was that of the 7-storey model. In hazard level 2, the highest energy absorption in Tabas earthquake was for the 15-storey model.

Participation rates of deformation and energy are compared in Figure 19.

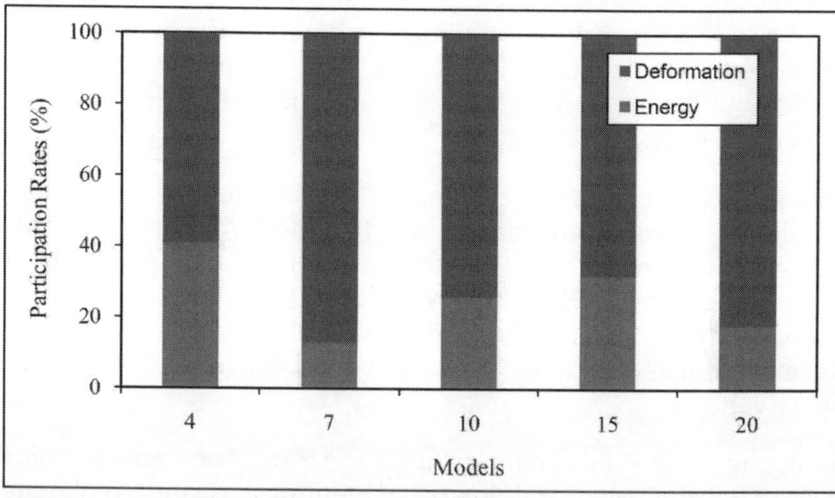

Figure 19 Participation rates of deformation and energy in damage.

Figure 19 illustrates the deformation as the most effective factor in the damage of buildings. Participation rates of deformation in damage of low-rise and high-rise buildings are 73% and 76%, respectively. Energy has 27% and 24% participation rates in damage.

CONCLUSIONS

In order to assess the reliability of structures subjected to ground motions, it is necessary to evaluate failure modes leading to cyclic deterioration. This paper has investigated the vulnerability of irregular buildings in near-field earthquakes using cumulative damage indices (ductility, energy, Park-Ang). In this study, the panel zone is considered, and irregular buildings are modeled three dimensionally. In order to do nonlinear dynamic analysis, each element was divided into ten parts. In these parts, the stresses and strains were derived from three points at the top, middle, and bottom corners of the section.

By increasing storeys in buildings, the value of required ductility in top storeys increased. Vulnerability of columns in the first storey in all buildings exceeded the allowable value. Deformation of columns was 93% in BSE-1 and 87% in BSE-2 which is higher than the beams in earthquakes.

In the most vulnerable storeys of low-rise and high-rise buildings, the values of dissipated energy are 34% and 26%, respectively. Severe damage and collapse occurred in the initial storeys of low-rise buildings and middle storeys of high-rise buildings due to seismic dissipated energy. For BSE-1 and BSE-2, energy dissipated in columns during earthquakes is 38% and 28% more than beams, respectively.

The maximum and minimum value of the Park-Ang damage index occurred in high-rise and low-rise buildings, respectively. The mean value of the Park-Ang damage index for beams in BSE-1 and BSE-2 was 0.75 and 1.26, respectively. This value was 0.55 and 1.1 for columns.

Participation rates of deformation and energy in damage of irregular low-rise buildings were 73% and 27%, respectively. Deformation and energy had 76% and 24% participation rates in damage of irregular high-rise buildings, respectively.

APPENDIX

Iranian seismic code

Every decade or so, a major earthquake strikes Iran. The 2003 Bam earthquake was the last major earthquake in Iran. Traditional buildings in Iran, especially in the rural areas, have very little resistance to earthquakes of higher magnitude. After numerous major earthquakes, in particular that of 1963 in Bouein Zahra and

1978 in Tabas, the Iranian government began the upgrading of the code of practice for earthquake protection. The Iranian Building and Housing Research Centre further revised the code, construction and design, and the updated and revised Iranian code for seismic-resistant design.

The design details of buildings according to the Iranian Seismic Code 2800 and the different seismic regions are categorized based on relevant seismic hazard analyses. As per Iranian Seismic Code 2800, earthquake lateral forces can be calculated using the following methods depending on the structure:

- Equivalent static analysis method
- Dynamic analysis method

An equivalent static analysis method is allowed just for regular structures which are not taller than 50 m or not irregular ones which are taller than 18 m. Other types of buildings must be designed using the dynamic analysis method. Since in this study the structures are irregular and are taller than 18 m, a dynamic analysis method is used in the design of buildings (Building and Housing Research Center 19992005; Asgarian et al. 2010; Gioncu and Mazzolani 2011).

AUTHORS' INFORMATION

MG is an assistant professor at the Faculty of Civil Engineering in Semnan University. YS is an MSc in Earthquake Engineering from Semnan University. AS is a PhD candidate in Earthquake Engineering at Semnan University.

COMPETING INTERESTS

The authors declare that they have no competing interests.

AUTHORS' CONTRIBUTIONS

MS defined the main theme and objective of the study and explained the research methodology. YS carried out the modeling and analysis, obtained the numerical results which are presented in the research work, and prepared the first draft of the paper. AS controlled the obtained numerical results, edited and finalized the manuscript, and participated in the global revision and discussion of the review performed. All authors read and approved the final manuscript.

REFERENCES

1. AISC: *Seismic provisions for structural steel buildings*. Chicago: Standard ANSI/AISC. American Institute of Steel Construction; 2005.

2. Alavi B, Krawinkler H: **Behavior of moment-resisting frame structures subjected to near-fault ground motions.** *J Earthquake Eng Struct Dynam* 2004,**33**(6):65–86.

3. ASCE/SEI 7–05: *Minimum design loads for buildings and other structures*. Reston: American Society of Civil Engineers; 2005.

4. Asgarian B, Sadrinezhad A, Alanjari P: **Seismic performance evaluation of steel moment resisting frames through incremental dynamic analysis.** *J Constr Steel Res*2010,**66**(2):178–90. 10.1016/j.jcsr.2009.09.001

5. Bray JD, Rodriguez-Marek A: **Characterization of forward-directivity ground motions in the near-fault region.** *Int J Soil Dynam Earthquake Eng* 2004,**24**(11):815–828. 10.1016/j.soildyn.2004.05.001

6. Brun M, Reynouard JM, Jezequel L, Ile N: **Damaging potential of low-magnitude near field earthquakes on low-rise shear walls.** *Int J Soil Dynam Earthquake Eng*2004,**24**(8):587–603. 10.1016/j.soildyn.2004.03.004

7. Bojórquez E, Reyes-Salazar A, Terán-Gilmore A, Ruiz SE: **Energy-based damage index for steel structures.** *J Steel Compos Struct* 2010,**10**(4):343–360.

8. Elnashai AS: **Assessment of seismic vulnerability of structures.** *J Constr Steel Res*2006,**62**(11):1134–1147. 10.1016/j.jcsr.2006.06.024

9. FEMA 356: *"Prestandard and Commentary for the Seismic Rehabilitation of Buildings", prepared by the American Society of Civil Engineers for the Federal Emergency Management Agency*. Washington, DC: FEMA; 2000.

10. Foutch DA, Yun SY: **Modeling of steel moment frames for seismic loads.** *J Constr Steel Res* 2002,**58**(5):529–564. 10.1016/S0143-974X(01)00078-5

11. Gerami M, Sivandi-Pour A: *"Introduction of earthquake loads amplification factor in design of seismic resistant steel buildings and comparison in various regulations". In: 1st national conference on retrofitting, Yazd, Iran (in Persian)*. 2008. 26–28 August 2008 26–28 August 2008

12. Gerami M, Sivandi-Pour A, Pourmeydani E: *"Assessment of effect of earthquake vertical component on seismic performance of ductile steel seismic resistant systems". In: 9th international congress on civil engineering, Isfahan, Iran (in Persian)*. 2012. 8–10 May 2012 8–10 May 2012

13. Gioncu V: **Ductility criteria for steel structures.** *J Constr Steel Res* 1998,**46**(1–3):443–444.

14. Gioncu V, Mazzolani FM: *Earthquake engineering for structural design*. London and New York: Spon Press; 2011.

15. Building and Housing Research Center: *Iranian code of practice for seismic resistant design of buildings, vol 2800*. 2nd edition. Tehran: Building and Housing Research Center; 1999.

16. Building and Housing Research Center: *Iranian code of practice for seismic resistant design of buildings, vol 2800*. 3rd

edition. Tehran: Building and Housing Research Center; 2005.

17. Kalkan E, Kunnath SH: **Assessment of current nonlinear static procedures for seismic evaluation of buildings.** *J Eng Struct* 2007,**29**(3):305–316. 10.1016/j.engstruct.2006.04.012

18. Karimi M, Hosseinzadeh N, Hosseini F, Kazem N, Kazem H: **Seismic vulnerability evaluation of pipe rack supporting structures in a petrochemical complex.** *Int J Adv Struct Eng* 2011,**3**(1):111–120.

19. Kim KD, Engelhardt MD: **Monotonic and cyclic loading models for panel zones in steel moment frames.** *J Constr Steel Res* 2002,**58**(5–8):605–35.

20. Krawinkler H: *The state-of-the-art report on system performance of moment resisting steel frames subjected to earthquake ground shaking, FEMA 355c.* Washington, DC: Federal Emergency Management Agency; 2000.

21. Kunnath SK, Chai YH: **Cumulative damage-based inelastic cyclic demand spectrum.** *J Earthquake Eng Struct Dynamics* 2004,**33**(4):499–520. 10.1002/eqe.363

22. Li S, Xie L: **Progress and trend on near field problems in civil engineering.** *Acta Seismologica Sinica* 2007,**20**(1):105–114. 10.1007/s11589-007-0105-0

23. Lignos D: *Sidesway collapse of deteriorating structural systems under seismic excitations.* 2008. Ph.D. thesis. Stanford University Ph.D. thesis. Stanford University

24. Mazzoni S, McKenna F, Scott MH, Fenves GL: *The Open System for Earthquake Engineering Simulation (OpenSEES) user command-language manual.* Berkeley: Pacific Earthquake Eng. Research Center, University of Califorinia; 2006.

25. Monavari B, Massumi A: **Estimating displacement demand in reinforced concrete frames using some failure criteria.** *Int J Adv Struct Eng* 2012.,**4**(4): 10.1186/2008-6695-4-4

26. Nethercot DA: **Design of building structures to improve their resistance to progressive collapse.** *J procedia Eng* 2011, **14:** 1–13.

27. Park YJ, Ang AH-S, Wen YK: *"Seismic damage analysis and damage-limiting design of RC buildings". Structural Research Series No.516*. Urbana: University of Illinois; 1984.

28. Pitocco M: **Information technology and management of diagnostics for analysis of seismic vulnerability in buildings.** *J Procedia Comput Sci* 2011, **3:** 352–360.

29. Poljanšek K, Fajfar P: *"A new damage model for the seismic damage assessment of reinforced concrete structures". In: 14th world conference on earthquake engineering, Beijing, China*. 2008. 12–17 October 2008 12–17 October 2008

30. Shibin L, Lili X, Maosheng G, Ming L: **Performance-based methodology for assessing seismic vulnerability and capacity of buildings.** *J Earthq Eng Vib* 2010,**9**(2):157–165.

31. Tirca L-D, Foti D, Diaferio M: **Response of middle-rise steel frames with and without passive dampers to near field ground motions.** *J Eng Struct* 2003,**25**(2):169–179.

9

CHAPTER

A Simplified Method in Comparison with Comprehensive Interaction Incremental Dynamic Analysis to Assess Seismic Performance of Jacket-type Offshore Platforms

M. R. Zolfaghari, A. Ajamy and B. Asgarian

Department of Civil Engineering, K.N. Toosi University of Technology, Tehran, Iran

ABSTRACT

The primary goal of seismic reassessment procedures in oil platform codes is to determine the reliability of a platform under extreme earthquake loading. Therefore, in this paper, a simplified method is proposed to assess seismic performance of existing jacket-type offshore platforms (JTOP) in regions ranging from near-elastic to global collapse. The simplified method curve exploits well agreement between static pushover (SPO) curve and the entire summarized interaction incremental dynamic analysis (CI-IDA) curve of the platform. Although the CI-IDA method offers better understanding and better modelling of the phenomenon, it is a time-consuming and challenging task. To overcome the challenges, the simplified procedure, a fast and accurate approach, is introduced based on SPO analysis. Then, an existing JTOP in the Persian Gulf is presented to illustrate the procedure, and finally a comparison is made between the simplified method and CI-IDA results. The simplified method is very informative and practical for current engineering purposes. It is able to predict seismic performance elasticity to global dynamic instability with reasonable accuracy and little computational effort.

Keywords: Uncertainty Ground motion records Comprehensive interaction incremental dynamic analysis (CI-IDA)Static pushover analysis (SPO)Jacket-type offshore platform

INTRODUCTION

The levels of damage resulted from some major events such as the San Fernando (1971), Loma Prieta (1989) and Northridge (1994) earthquakes indicate that it is essential to modify the building codes to improve seismic resistance in structures. Therefore, the Performance-Based Earthquake Engineering (PBEE) methodology has been applied in design and construction from 2000 to evaluate the seismic performance of existing structures under extreme loads (Krawinkler and Miranda 2004). While the ever-increasing processing power of computers has improved upon analytical applications of structural models, the analyses have been conducted from elastic static analyses to dynamic elastic, nonlinear static and finally nonlinear dynamic analyses.

The nonlinear static analysis normally called pushover analysis is a procedure in which a structural model is subjected to a predetermined lateral load pattern and demonstrates the relative inertia forces generated at locations of substantial mass. In this procedure, the intensity of the load increases (the structure is 'pushed') and incremental process continues until a predetermined displacement is achieved. Among several pushover analyses, the static pushover (SPO) analysis has no strict theoretical base such that the response of the structure is controlled by the first mode of vibration through introducing a constant load pattern. Moreover, the principle objective of SPO is based on structural static theory and it does not correctly predict the dynamic failure modes (Krawinkler 2001).

Since the result of the static pushover provides information that is not obtained from an elastic static and dynamic analysis, it is one of the most important methods to offer seismic performance. The following are the advantages of SPO analysis (Govind et al. 2014):

1. The method shows overall structural behaviours and their distributions along the height.

2. It provides the sequential formation in the individual structural components, and also identifies location of weak points or potential failure modes in the structure.

3. It shows consequences of strength deterioration on the behaviour of structural system in both plan and elevation.

In addition, there are some limitations in the SPO analysis described in the following (Khan and Vyawahare 2013):

1. This method represents inaccurate responses in which higher mode effects are significant.

2. It will be necessary to perform the analysis with displacement rather than force control.

3. SPO analysis neglects duration effects, number of stress reversals and cumulative energy dissipation demand in the structures.

4. This method cannot consider the progressive changes in modal properties.

5. Finally, the SPO method is able to detect only the first local mechanism.

Chopra and Goel (2001) suggested the modal pushover analysis (MPA) method based on structural dynamics theory. While the MPA method provides sufficient accuracy in the estimation of seismic demands, it is conceptually simple and straightforward (Chopra and Goel2001). Moreover, Chopra et al. (2004) proposed the modified modal pushover analysis method (MMPA) approach assuming higher modes result in only elastic behaviour of the structure (Chopra et al. 2004). Since the MMPA method neglects the variability of higher modes that can lead to nonlinear behaviours stage in structures, it cannot always predict accurately structural behaviours (Themelis 2008).

To remedy this deficiency in the procedure, Vamvatsikos (2002) proposed the incremental dynamic analysis (IDA) method. This approach is an emerging structural analysis method performing a series of nonlinear dynamic analyses under a suite of multiply scaled ground motion records. The method is usually very complex and time consuming because of the complex nature of strong ground motions.

To reassess the jacket offshore platforms, there are two general forms: (1) static pushover analysis (2) time domain analysis. Asgarian and Ajamy (2010) applied the IDA method for new designed jacket-type offshore platforms in the Persian Gulf and defined different structural limit states. Also, Asgarian and Rahman Shokrgozar (2013) evaluated the seismic performance of an existing jacket-type offshore platform with float-over deck using a probabilistic method. They estimated the mean annual frequency and confidence levels in this type of structures.

Once the IDA procedure is applied to perform nonlinear dynamic analyses, aleatory (record-to-record) uncertainty only is considered. Obviously, the best prediction will be achieved when the effects of epistemic uncertainty are also considered. Therefore, Cornell et al. (2002) and Ellingwood (2007) proposed one approach to combine the effects of different sources of uncertainties in terms of the confidence interval approach. The distribution associated with epistemic uncertainty in this case may be obtained from the first-order-second-moment reliability, Monte Carlo Simulation (MCS) methods or expert judgment.

Liel (2008) used the response surface method as a functional relationship between the input random variables and limit state criterion to find collapse capacity limit. She assessed collapse risk of reinforcedconcrete moment frame buildings. Moreover, the MCS method along with response surface was applied to propagate modelling uncertainties and fit a response surface to sensitivity analysis results. Furthermore, Liel (2008) have shown that in very nonlinear structural responses, the effect of assumed epistemic uncertainty can be more significant in the assessment.

In oil and gas industry, the effects of epistemic uncertainty have been studied by Ajamy et al. (2014) in an existing jacket-type offshore platform. They proposed a new method named the Comprehensive interaction incremental dynamic analysis (CI-IDA) method to propagate epistemic uncertainties associated with aleatory uncertainty in different parts of structural systems such as surrounding soil, pile, structural elements and finally ground motion records.

Since accurate estimates of the seismic performance require performing the costly analyses, a simplified method is introduced in this paper. It creates a direct connection between static pushover analysis and time domain analyses. Then, the sufficiency of the simplified method is examined in an existing jacket-type offshore platform in the Persian Gulf using comparison with CI-IDA results.

REQUIREMENTS TO PERFORM THE SIMPLIFIED METHOD

The simplified procedure includes three steps: (1) to create a structural model to offer realistic situations (2) to perform SPO analysis and generate SPO curves in different levels (3) to calculate different slops in the SPO curve.

Jacket-type offshore platform modelling

Since oil and gas industry relies on offshore platforms, the reassessment of jacket-type offshore platforms under extreme loads is of significant importance. In this paper, one of the existing jacket-type offshore platforms constructed in the early 1970s has been chosen to reassess seismic performance.

The platform is in the Persian Gulf, approximately 100 km southwest of the Kharg Island export terminal. The field is on the Iran-Saudi Arabian water border, with the Saudi-Arabian portion of the field being called the Marjan field.

Figure 1 presents the two-dimensional model of the selected offshore platform. As shown in Fig. 1, the platform is a six-legged platform with 75 m height. To illustrate the methodology, a two-dimensional model associated with its deep piles was created in the OpenSees (2006) analysis platform.

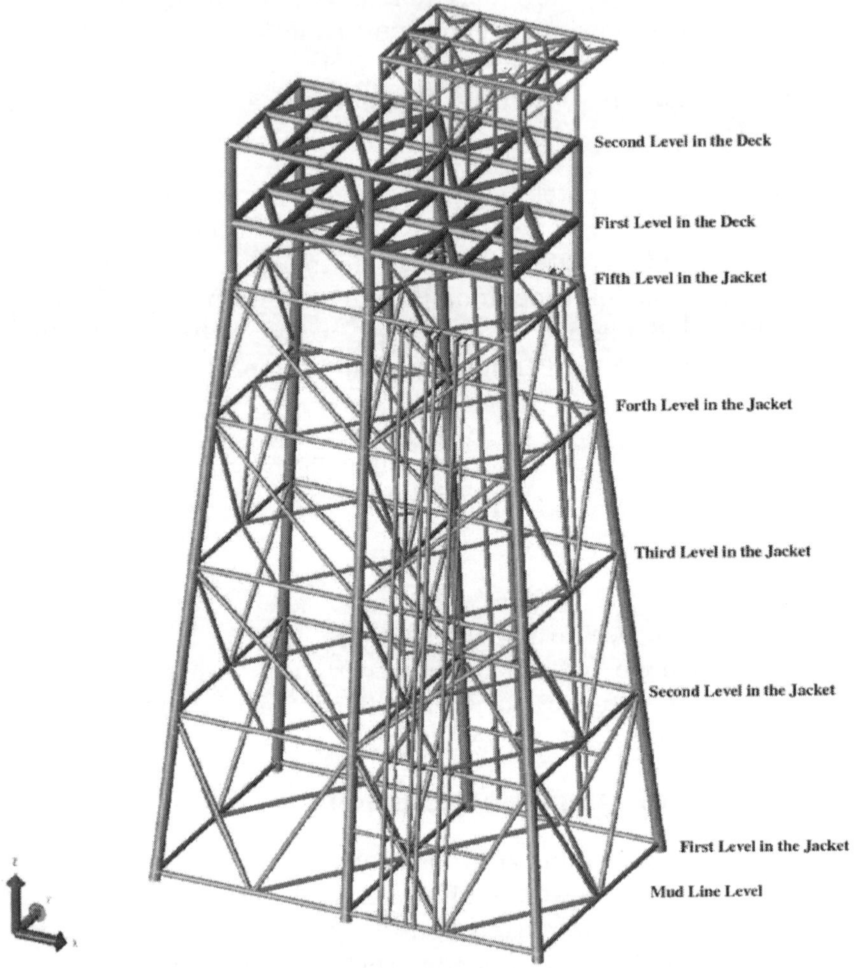

Fig. 1 A view of the two-dimensional model configuration

The main structural subsystems of the model consisted of 42 frame elements which include 3 jacket leg members, 25 beam members, 14 diagonal brace members. The platform geometry includes the total length and the number of spans of the jacket and the deck, the cross-sectional dimensions of the piles, legs, beams and braces and support details to model boundary conditions. Moreover, the model entails conductors and risers and their dimensions are based on the existing data in the drawings.

In simulation part, all deck members were modelled by elastic members and a forceBeamColumn element object was used to simulate the jacket members. The object is based on the iterative force-based formulation and addresses the nonlinear behaviours in terms of both distributed plasticity and plastic hinge integration. Moreover, to simulate pile elements in OpenSees software, a dispBeamColumn element object which is a distributed-plasticity and displacement-based beam-column element was used.

The cross-sections of the elements were modelled using the fibre element to model the nonlinear behaviours of the platform. In this research, the cross-sectional geometric properties of tubular members consist of outer diameter and thickness as exists in the drawings. The application of fibre elements is discussed in more detail elsewhere (Asgarian et al. 2005, 2006) to address buckling and post-buckling behaviours of tubular members and also nonlinear analyses of platforms.

To create an exact geometric transformation, Corotational Transformation was applied in the jacket elements. It uses beam stiffness and resisting force in the basic system to convert the global coordinate system. In piles, P-Delta Transformation was used to consider the geometric nonlinearities and large deformation effects in the nonlinear time history analyses.

The mass of the platform consists of the distributed mass of piles, legs, beams and braces considered in the length of modelled elements. Based on the reported masses in the in-place analysis report, the mass of various appurtenances such as conductors, risers, mud mats, boat landings, walkways, and caissons is added to the mass of the electrical, mechanical and structural parts; then, it is considered as lumped mass at the jacket structural nodes. In this study, the presence of water is neglected and the effect of water–soil–pile–structure interaction is not addressed while the model includes flooded members.

Since structure, pile and surrounding soil interact with each other under static and dynamic loading, their interaction is complex and requires a comprehensive understanding. On the other hand, this issue plays a significant role in determining the response of structures such as offshore platforms during an earthquake. For instance, seismic soil–pile–structure interaction (SSPSI) usually increases the periods of the system or varies the system damping (Azarbakht et al. 2008).

The several researchers developed numerous simplified methods to address the effects of SSPSI. Among these models, those based on the static/dynamic beam-on-nonlinear-Winkler-foundation (BNWF) method, often referred to as the p–y method, are commonly utilized to simulate SSPSI problems (Winkler 1867).

In this study, the BNWF method was used to model the effects of SSPI. In this method, independent horizontal and vertical nonlinear springs have been applied along the pile shaft to represent thicknesses, stiffness, and damping characteristics of each layer. The layers (39 layers) were modelled using p–y, t–z and q–z elements according to API-RP2A (1993) provisions and then attached to pile nodes through zero-length elements.

In the analysis part, the gravity analysis is carried out using Newton algorithm. The algorithm solves the nonlinear equations and is able to be updated each iteration. Moreover, the Norm Displacement Increment test command was used to construct a Convergence test object. The command determines if convergence has been achieved at the end of an iteration step.

In time history analyses, the earthquake excitation is simulated using MultipeSupport Excitation pattern. It constructs an ImposedMotionSP constraint used to enforce the response of a degree-of-freedom at a node in the model. The response is obtained from the GroundMotion object associated with the constraint. In the material part, the uniaxialMaterial Steel01 command was used to consider material properties. It applies to construct a uniaxial bilinear steel material object with kinematic hardening and optional isotropic hardening described by a nonlinear evolution equation.

Since the dynamic characteristics of the offshore platform are explicitly portrayed through modal analysis, it was carried out in OpenSees Post Processor (OSP) before each transient analysis to represent natural frequencies. Then, the results were compared with the periods computed during the assessment stage using the Structural Analysis Computer System (SACS) software.

The frequencies and the mode shapes assumed by the platform are determined analytically based on the stiffness, mass, and damping properties of the simulated system. The results in the OSP software match well with the SACS results. For instance, the first natural period is equal to 2.17 s in OSP while it is equal to 2.25 s in SACS. Also, the mode shape of the periods corresponds to motion in the transverse direction of the platform.

The modal results, specifically modal periods, are the main parameters used in time history analyses. Moreover, geometric nonlinearity was included in all dynamic analyses and damping was also considered using Rayleigh damping. Since the transient analysis was often unable to converge to a solution, a solution procedure script was developed within the analysis file that tried a number of different solution algorithms, time steps, and convergence criterion until a solution could be achieved.

STATIC PUSHOVER ANALYSIS

SPO analysis is applied to establish the static ultimate strength of the platform and the failure sequence for the selected loading pattern. Moreover, it is utilized to monitor seismic performances of the platform.

In this study, Fig. 2 shows the variation of base shear versus maximum inter-level drift in the SPO curve at the mud line level of the offshore platform. Based on FEMA 350 (2000a, b), ASCE 41-06 (2007) and engineering judgment, the collapse prevention (CP) is considered by the value of $\theta_{max} = 2\%$ in jacket-type offshore platforms. In the Fig., there are three regions including the elastic linear, collapse prevention and first failure regions in the SPO curve.

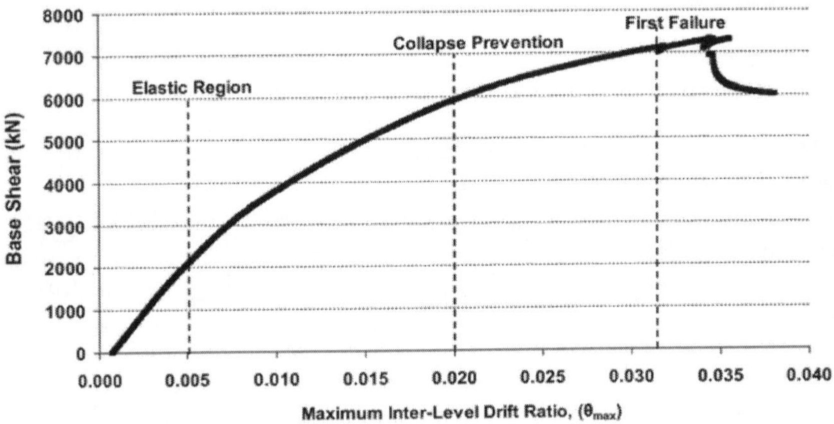

Fig. 2 The static pushover curve at the mud line level

Since the SPO curve is usually based on base shear versus $\theta_{\max}$ coordinates, it needs to be transformed into IM and DM axes. Therefore, it is necessary to divide the base shear by the total structural mass by times a proper factor to convert it to appropriate intensity measure. Figure 3 represents the SPO curves at the working point and mud line levels of the jacket and the top level of the deck. They show the variability of $\theta_{\max}$ versus $S_a(T_1, 5\%)$ axes to offer the damage levels in the platform. It is obvious that the most vulnerable level is the mud line level.

The simplified method using SPO analysis

To introduce the simplified approach, it is necessary to obtain the slopes of the SPO curve in different parts. Figure 4 represents the SPO curve at the mud line level based on DM versus IM axes. As illustrated in the Fig., since there are three separate regions in the SPO curve, it is necessary to obtain three different slopes to accomplish the simplified method.

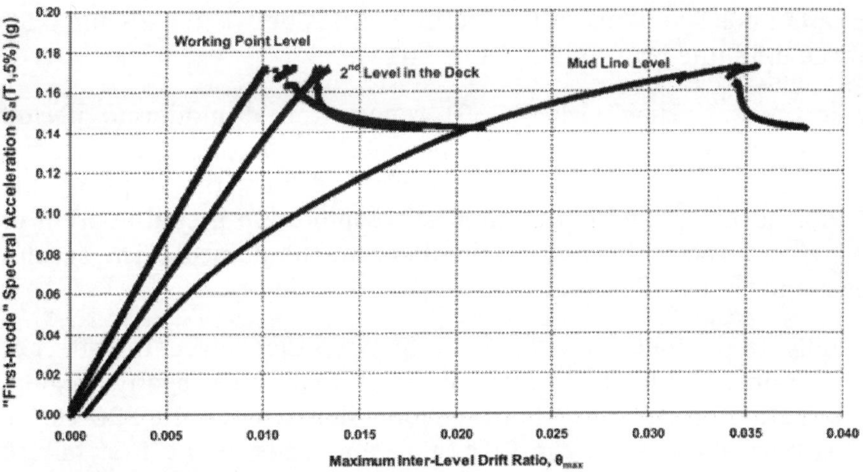

Fig. 3 The SPO curves at the working point and mud line levels of the jacket and 2nd level of the deck

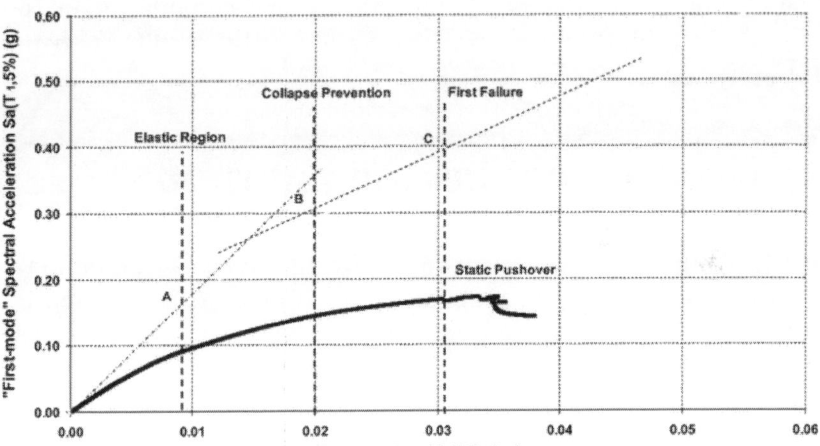

Fig. 4 The simplified curve (*dashed line*) and the SPO curve (*solid line*) at the mud line level

At the beginning, to define the elastic linear region, the elastic segment of the simplified curve coincides the elastic slope of the SPO curve; then, the effect of uncertainty is gradually added towards the elastic slope in the SPO curve. In this place, the point A is obtained. The effect of uncertainty varies between 0.35 and 0.50 at the mud line level (Ajamy et al. 2014).

Veletsos and Newmark (1960) reported that moderate period structures follow "equal-displacement" rule in non-negative stiffness region. Therefore, in the second part, the slope of this region in the simplified method corresponds to a continuation of the elastic region in the simplified curve (the point B). The second part ends at the CP region.

Finally, to define the third part in the simplified curve, the effect of uncertainty is added the slope of the SPO curve in negative region. It should be mentioned that the slope decreases in the SPO curve. This part starts from the CP region and ends at the first failure region (the point C).

The aforementioned steps have been illustrated in Fig. 4 from the results of SPO curve. In the Fig., the dashed line shows the simplified curve and the solid line shows the SPO curve at the mud line level. Since the effects of uncertainty are considered in the simplified method, it is necessary to validate the results with the CI-IDA method.

FUNDAMENTALS OF THE CI-IDA METHOD

The CI-IDA method involves three following steps: (1) to simulate structural models to present realistic modelling case, (2) to identify uncertainties in terms of aleatory and epistemic uncertainties, (3) to carry out nonlinear dynamic analyses subjected to ground motion records and to provide the relationship between damage measure (DM) and intensity measure (IM) using the curves of single-CI-IDAs and multi-record CI-IDAs and their summary.

Uncertainty considered in this study

In this paper, it is assumed independence between the aleatory and epistemic random variables, as illustrated in Fig. 5. Epistemic uncertainties are defined in different parts of the structural system including pile, structural elements and surrounding soil and

record-to-record variability in actual ground motion records is treated as aleatory uncertainty.

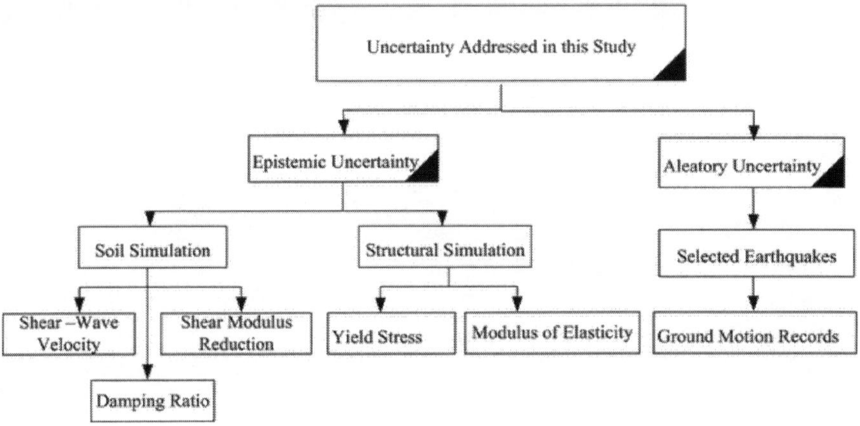

Fig. 5 Uncertainty addressed in this Study

In the structural simulation section, yield stress of legs and braces and modulus of elasticity and in the soil simulation section, shear-wave velocity, shear modulus reduction and damping ratio are considered as epistemic uncertainty. Table 1 shows statistical data of the structural and soil properties in different parts including the used symbol, the type of distribution function, mean values of the variables, their coefficient of variation (COV) and finally the used references.

Table 1 Statistical data of the structural and soil properties

Random variables	Symbol	Type	Mean	COV	References
Parameters influencing uncertainties in structural element model					
Yield stress of legs	$f_{Y,L}$	Lognormal	360 Mpa, 355 Mpa	0.07	Haselton (2006); JCSS (2001)
Yield stress of braces	$f_{Y,B}$	Lognormal	360 Mpa, 355 Mpa	0.07	Haselton (2006); JCSS (2001)
Modulus of elasticity	E_S	Lognormal	2.00×10^5 Mpa	0.03	Haselton (2006); JCSS (2001)
Parameters influencing uncertainties in surrounding soil and pile element model					
Shear–wave velocity	V_S	Lognormal	Computed	Computed	Toro (1995)
Shear modulus reduction	G/G_{max}	Normal	Computed	Computed	Darendeli and Stokoe (2001)
Damping ratio	D	Normal	Computed	Computed	Darendeli and Stokoe (2001)

The technique of latin hypercube sampling (LHS) in conjunction with simulated annealing (SA) optimization (Vorechovsky 2004) was applied to propagate the effects of epistemic uncertainties in the model. The procedure is described in many references elsewhere (Vorechovsky 2004,2012; Vorechovsky and Novak 2009) and only a brief summary is provided here.

In this technique, a difference matrix named the error correlation matrix (E) is determined according to Eq. 1. It is the difference between the target correlation matrix (*T*) and the actual correlation matrix (*A*). *T* is introduced by user and A can be estimated by Pearson's correlation coefficient, Spearman's rho or Kendall's tau (Vorechovsky and Novak 2009)

$$E = T - A. \tag{1}$$

In the equation, the number of the input random variables is equal to five (based on the Table 1) and the number of simulations is assumed equal to 15 (Vorechovsky 2012). After the application of LHS associated with SA, the following matrix (2) is presented in which the lower triangle is equal to the target correlation matrix \(T\) (*T*) and the upper triangle is equal to the Spearman rank order correlation matrix \(A\) (*A*). The correlation coefficients of the matrix reflect the studied epistemic uncertainties in each simulation

$$\begin{pmatrix} f_y & E_S & V_S & \frac{G}{G_{max}} & D \\ 1 & -0.00714 & 0.01071 & 0.00000 & -0.01429 \\ 0 & 1 & -0.02500 & 0.00714 & -0.00714 \\ 0 & 0 & 1 & -0.00714 & 0.00357 \\ 0 & 0 & 0 & 1 & -0.48929 \\ 0 & 0 & 0 & -0.5 & 1 \end{pmatrix} \begin{matrix} f_y \\ E_S \\ V_S \\ \frac{G}{G_{max}} \\ D \end{matrix} \cdot \tag{2}$$

To address record-to-record variability according to Table 2, a set of 10 ground motion records is enough to be used to offer proper accuracy in the estimation of structural behaviours. There are two main factors affecting the record selection process. One of them is the soil type and the other is the average fundamental period of the studied structure (Shome and Cornell 1999; Azarbakht and Dolšek 2007).

Table 2 The set of ten ground motion records applied

No.	Station	Component	Mechanism	PGA
1	Parkfield—Fault Zone 14 in Coalinga-01	090	Reverse-oblique	0.274
2	Slack Canyon in Coalinga-01	045	Reverse-oblique	0.166
3	Parkfield—Fault Zone 16 in Coalinga-01	000	Reverse-oblique	0.195
4	25 SMART1 C00 in Taiwan SMART1	000	Reverse-oblique	0.172
5	APEEL 7—Pulgas in Loma Prieta	000	Reverse-oblique	0.156
6	Hayward—BART Sta in Loma Prieta	310	Reverse-oblique	0.156
7	Hollister Diff. Array in Loma Prieta	165	Reverse-oblique	0.269
8	Palo Alto—SLAC Lab in Loma Prieta	360	Reverse-oblique	0.278
9	LA—Univ. Hospital in Northridge-01	095	Reverse	0.214
10	Lake Hughes #12A in Northridge-01	090	Reverse	0.174

The ground motion records belong to the far-field ground motions from the PEER (2006) database. Based on the local geotechnical report, the site class D is considered according to recent NEHRP (2001) seismic provisions. Also, it includes large source-to-site distances between 20 and 70 km with relatively large magnitudes of 6.0–7.3, and reverse or reverse-oblique faulting mechanisms.

In addition to the above criteria, Arias intensity was used to select the records. Arias intensity is a ground motion parameter that captures the potential destructiveness of an earthquake as the integral of the square of the acceleration time history. It correlates well with several commonly used demand measures of structural performance, liquefaction, and seismic slope stability (Travasarou et al. 2003). Using site-specific hazard curve, spectral accelerations for 2 and 10 % probability of exceedance in 50 years are equal to 0.26 and 0.16 g, respectively (Ajamy et al.2014).

After the record selection process, each record was scaled in ten levels through the hunt and fill algorithm (Vamvatsikos and Cornell 2004). To complete the requirements in CI-IDA, the 5 %-damped first-mode spectral acceleration of the platform $(S_a(T_1, 5\,\%))$ and maximum inter-level drift ratio (θ_{max}) were chosen as a relative efficient IM and DM, respectively.

Since the seismic response of the surrounding soil strongly affects the structural performance, equivalent-linear site response analyses are used to simulate the effects of the ground response using one-dimensional (1D) models. The models assume that seismic waves propagate vertically through horizontal sediment layers. Among several computer programs, the DEEPSOIL software was selected to perform the seismic site response analysis (Hashash et al. 2012).

According to Table 1, treatment of uncertainties is considered in the input ground motions, the nonlinear properties of the surrounding soil and the shear-wave velocity of the site (Rathje et al.2010). DEEPSOIL uses outcropping motions as the input ground motions, and then the nonlinear properties and the shear-wave velocity are defined in the calculation part by user.

To simulate the shear-wave velocity, the statistical models of Toro (1995) were used in which a lognormal distribution at mid-depth of the layer and an interlayer correlation are applied. Equation (2) predicts the shear-wave velocity in the *i*th layer $(V_s(i))$.

$$V_s(i) = \exp\{\ln[V_{\text{median}}(i)] + Z_i \cdot \sigma_{\ln V_s}\} \quad (3)$$

where $V_{\text{median}}(i)$ is the mean shear-wave velocity at the mid-depth of the layer $\sigma_{\ln V_s}$ is the standard deviation of the natural logarithm of $V_s(i)$, Z_i is the standard normal variable of the *i*th layer. All random variables are generated based on the site class and defined in more detail elsewhere (Toro1995).

To model the nonlinear soil property, the empirical models of Darendeli and Stokoe were used (JCSS 2001). In these models, the mean variation of the nonlinear soil properties follows a normal distribution; the standard deviation of the normalized shear modulus (σ_{NG}) and the standard deviation of the damping ratio (σ_D) are predicted by Eqs. (4, 5).

$$\sigma_{NG} = 0.015 + 0.16\sqrt{0.25 - (G/G_{\max} - 0.5)^2} \quad (4)$$

$$\sigma_D = 0.0067 + 0.78\sqrt{D(\%)} \tag{5}$$

The variation of Eqs. 3 and 4 is described in detail elsewhere (33). Since G/G_{max} and D curves are not independent $(\rho_{D,NG} < 0)$, they are correlated G/G_{max} and D curves from baseline (mean) curves generated by Eqs. 6 and 7 for each shear strain value γ.

$$G/G_{max}(\gamma) = [G/G_{max}(\gamma)]_{mean} + \varepsilon_1 \times \sigma_{NG} \tag{6}$$

$$\begin{aligned} D(\gamma) = [D(\gamma)]_{mean} + \rho_{D,NG} \times \varepsilon_1 \times \sigma_D + \sigma_D \\ \times \sqrt{\left(1 - \rho_{D,NG}^2\right)} \times \varepsilon_2 \end{aligned} \tag{7}$$

where ε_1 and ε_2 are uncorrelated normal random variables with zero mean and unit standard deviation $[G/G_{max}(\gamma)]_{mean}$ and $[D(\gamma)]_{mean}$ are the baseline values evaluated at strain level γ; σ_{NG} and σ_D are the standard deviations computed from Eqs. (3) and (4) at the baseline values of $[G/G_{max}(\gamma)]_{mean}$ and $[D(\gamma)]_{mean}$, respectively, and $\rho_{D,NG}$ is the correlation coefficient between G/G_{max} and D.

Comprehensive interaction incremental dynamic analysis

CI-IDA is performed by conducting a series of nonlinear time history analyses. They require several months to accomplish the whole process. Once IM is incrementally increased in each analysis, a DM, such as global drift ratio, is monitored during each analysis. Finally, the extreme values of a DM are plotted against the corresponding value of the ground motion IM for each intensity level to produce a single-CI-IDA curve for the platform and the chosen earthquake record.

Since a single-CI-IDA curve is not able to fully capture the seismic performance, a collection of single-record CI-IDA curves is generated in numerous sets under different accelerograms. For instance, Fig. 6 is multi-record CI-IDA curves and represents the variation of maximum inter-level drift ratio $(\theta_{\max})$ versus the 5 %-damped first-mode spectral acceleration $(S_a(T_1, 5\%))$ at 3rd level of the jacket using the 150 raw single-record CI-IDA curves. In the Fig., the curves exhibit maximum inter-level drift ratios change from 0.12 to 0.245, 0.15 to 0.295 and 0.205 to 0.505 % for the elastic linear region, the 10 and 2 % in 50 years ground motion levels, respectively.

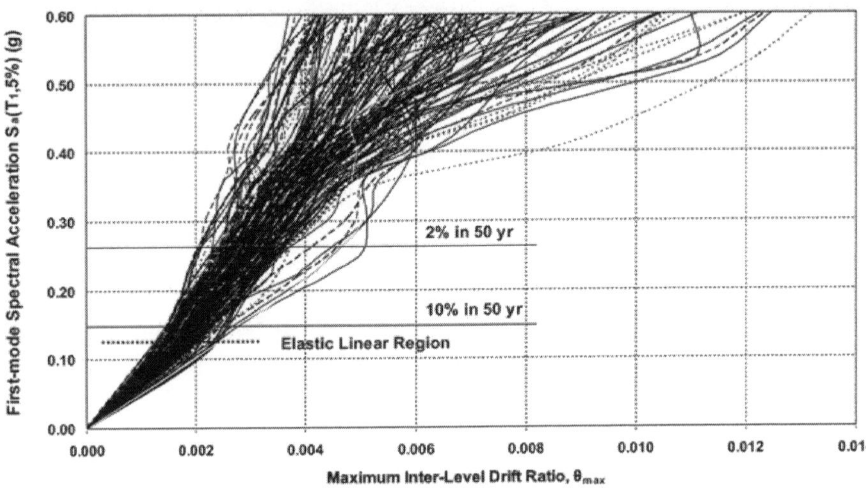

Fig. 6 The 150 raw single-record CI-IDA curves at 3rd level of the jacket

In multi-record CI-IDA curves, there is practical information that can be provided using appropriate summarization techniques. Therefore, the cross-sectional fractile technique was applied to generate mean, 16 and 84 % fractiles. Figure 7 shows multi-record CI-IDA curves, the 16, 50 and 84 % fractile CI-IDA curves at the working point level.

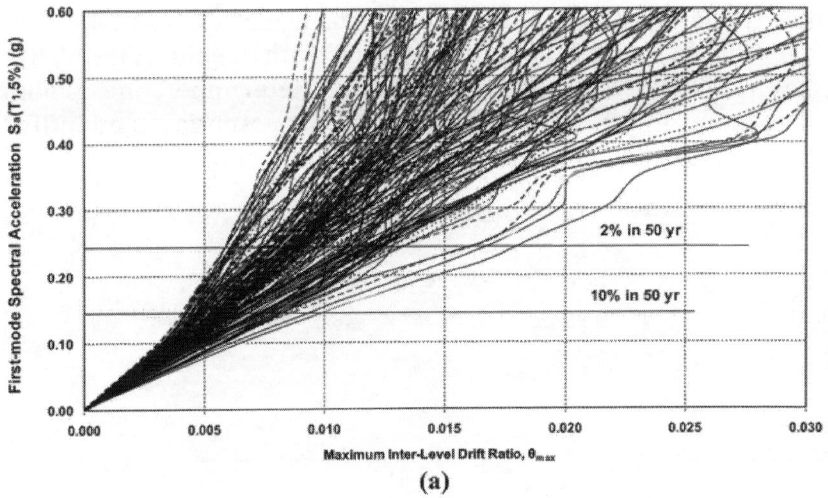

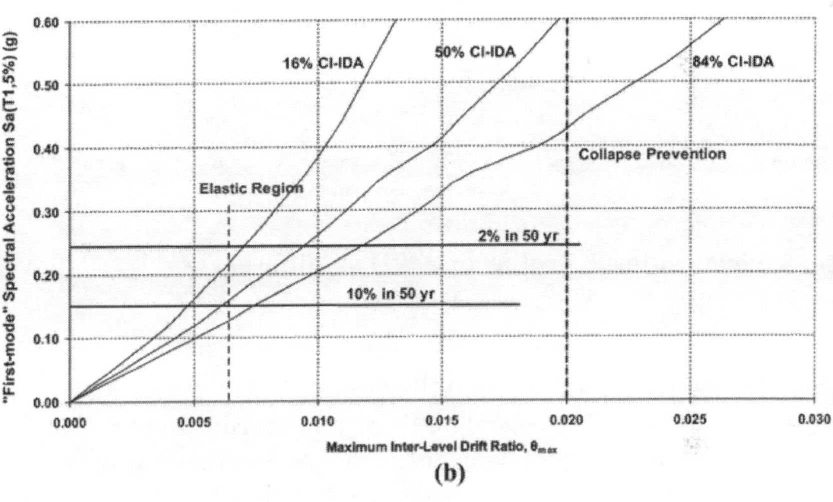

Fig. 7 Multi-record CI-IDA curves and their summaries at the working point level **a** the 150 raw single-record CI-IDA curves, **b** the 16, 50 and 84 % fractile CI-IDA curves

Moreover, Fig. 7 graphically defines some limit states at the working point level of the offshore platform. For instance, the 50 % CI-IDA curve presents θ_{max} equal to 0.65, 0.60 and 0.90 % for the elastic region, the 10 and 2 % in 50 years ground motion levels, respectively.

Since the jacket-type offshore platforms are as tall structures, each level presents different drift patterns with height (Figs. 7, 8). To assess seismic performances and determine the collapse mechanisms in the offshore platforms, it is essential to quantify the seismic structural performance in the whole platform.

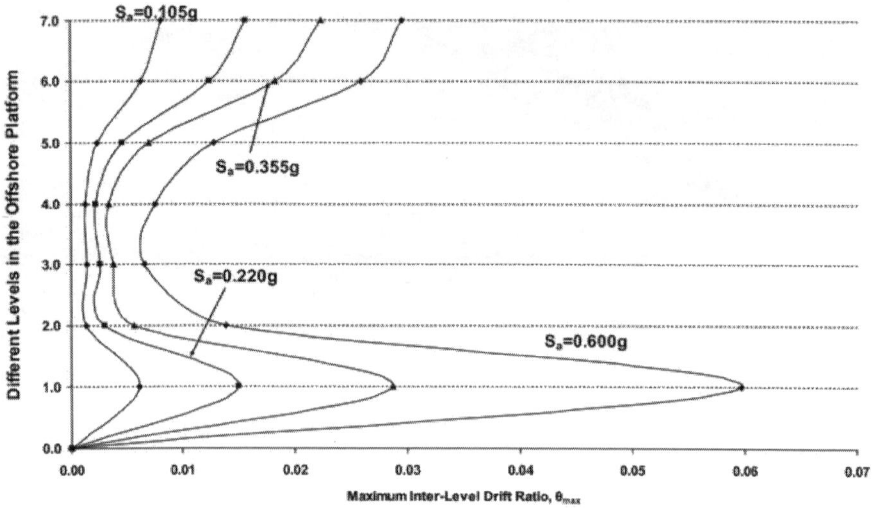

Fig. 8 Level-to-level profile of $\theta_{\max}$ at different $S_a(T_1, 5\,\%)$ levels in all records

Figure 8 displays level-to-level profile of $\theta_{\max}$ different $S_a(T_1, 5\,\%)$ levels of all selected records. The first level to the fifth level shows the levels of the jacket and the rest belongs to the deck. The figure shows there are irregularities in elevations such that the first level in the jacket acts as a weak level relative to stronger upper levels. Also, the variation of $\theta_{\max}$ reflects the importance of addressing the effects of epistemic and aleatory uncertainties, simultaneously, to predict the seismic performance.

On the other hand, Fig. 9 shows the median variations of $\theta_{\max}$ versus $S_a(T_1, 5\,\%)$ in each level subjected to all records, separately. As shown in Figs. 8 and 9, while the mud line level acts as a fuse in the whole platform, there is a relative rigid core in the second, third and fourth levels of the jacket and gradually the levels of damage increase in the fifth level of the jacket and the first and second levels of the deck. It should be mentioned that the mud line level elevation is very close to the first level of the jacket; therefore, they have very similar seismic performances.

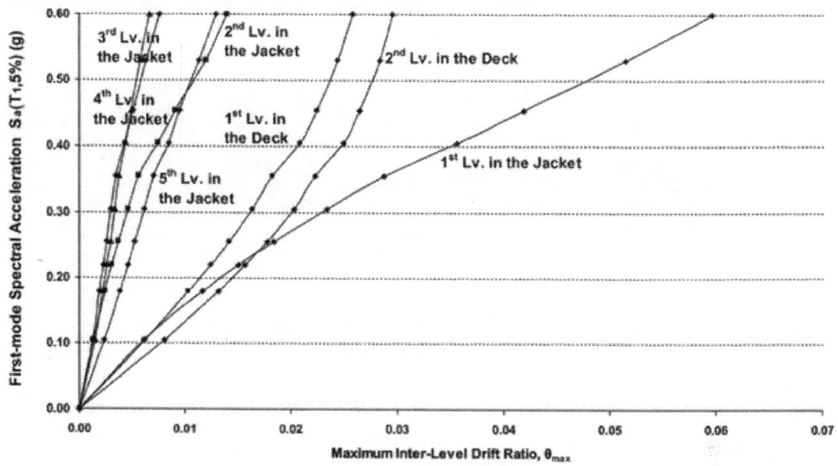

Fig. 9 The median CI-IDA curves of the offshore platform resulted from all records

Comparison of the simplified method and CI-IDA

In this study, the simplified and the CI-IDA methods were performed in the same offshore platform. Figure 10 represents comparison of the median CI-IDA curve (thick dash line), the simplified method curve (dash line) and the SPO curve (solid line) at the mud line level.

As per the results, the maximum inter-level drift ratio at the 10 % in 50 years ground motion level resulted from the simplified method and mean CI-IDA is about 0.8 % while at the 2 % in 50 years ground motion level is 1.4 and 1.7 %, respectively.

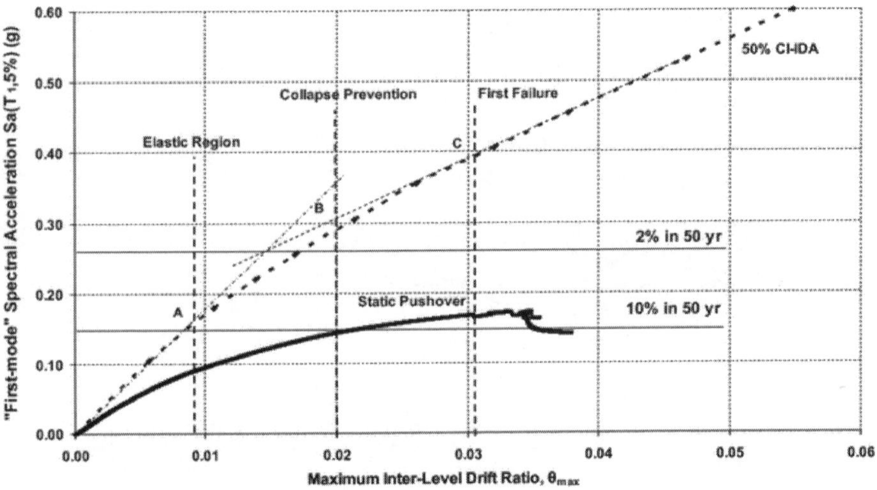

Fig. 10 Comparison of the median CI-IDA curve (*thickdash line*), the simplified method curve (*dash line*) and the SPO curve (*solid line*) at the mud line level

Moreover, θ_{max} is equal to 0.9 % in the elastic linear region of the simplified method and mean CI-IDA curves and ends at $S_a(T_1, 5\%) \approx 0.16\,g$ and $0.15\,g$, respectively. Furthermore, the CP in the simplified method and mean CI-IDA occurs at $S_a(T_1, 5\%) \approx 0.31\,g$ and $0.28\,g$, respectively. The results indicate that the simplified method is a straightforward approach that is able to predict different limit states with reasonable accuracy.

CONCLUSIONS

In this paper, a simplified method has been proposed. It can approximate the seismic demands of jacket-type offshore platforms from elasticity to global collapse. The method is based on the static pushover (SPO) and is able to predict the 50 % fractile of the comprehensive interaction incremental dynamic analysis (CI-IDA) curve for the platform with reasonable accuracy. Since CI-IDA is a

Proceedings of the 2nd World Conference on Earthquake Engineering, Tokyo, Japan, pp 895–912

36. Vorechovsky M (2004) Stochastic fracture mechanics and size effect, Ph.D. dissertation, the Brno University of Technology, Faculty of Civil Engineering, p 180

37. Vorechovsky M (2012) Correlation control in small-sample Monte Carlo type simulations II: analysis of estimation formulas, random correlation and perfect uncorrelatedness. J Probab Eng Mech, pp 105–120

38. Vorechovsky M, Novak D (2009) Correlation control in small-sample Monte Carlo type simulations I: a simulated annealing approach. J Probab Eng Mech 24:452–462

39. Winkler E (1867) Die Lehre von der Elasticitaet und Festigkeit. Prag, Dominicus

Index